# SpringerBriefs in Food, Health, and Nutrition

Springer Briefs in Food, Health, and Nutrition present concise summaries of cutting edge research and practical applications across a wide range of topics related to the field of food science.

**Editor-in-Chief**
Richard W. Hartel, *University of Wisconsin–Madison, USA*

**Associate Editors**
J. Peter Clark, *Consultant to the Process Industries, U*
John W. Finley, *Louisiana State University, USA*
David Rodriguez-Lazaro, *ITACyL, Spain*
David Topping, *CSIRO, Australia*

For further volumes:
http://www.springer.com/series/10203

Gloria Sánchez

# Hepatitis A Virus in Food

Detection and Inactivation Methods

Gloria Sánchez
IATA-CSIC, Novel Materials, Nanotechnology
Institute of Agrochemistry and Food Technology
Valencia, Spain

ISBN 978-1-4614-7103-5          ISBN 978-1-4614-7104-2 (eBook)
DOI 10.1007/978-1-4614-7104-2
Springer New York Heidelberg Dordrecht London

Library of Congress Control Number: 2013934549

Printed on acid-free paper

Springer is part of Springer Science+Business Media (www.springer.com)

# Abstract

Hepatitis A virus (HAV) is responsible for around half of the total number of hepatitis infections diagnosed worldwide. HAV infection is mainly propagated via the fecal-oral route, and as a consequence of globalization, transnational outbreaks of foodborne infections are reported with increasing frequency. Therefore, in this review, state-of-the-art information regarding molecular procedures for HAV detection in food as well as efficacy of common food-manufacturing processes is compiled. The purpose of this book is to consolidate basic information on various aspects of HAV and to provide guidelines for HAV prevention and control across the food supply chain from preharvest to manufacturing.

# Contents

# Chapter 1
# Introduction

Hepatitis A infection is the leading cause of viral hepatitis throughout the world. According to the World Health Organization, there are more than 1.4 million new cases of hepatitis A worldwide every year, and the cost of foodborne illness caused by hepatitis A in the USA is now estimated to be more than 36,000 US dollars, according to an analysis by Ohio State University (Scharff 2012).

Hepatitis A infection is caused by the hepatitis A virus (HAV), a small (27–32 nm), nonenveloped, single-strand RNA virus with an estimated infectious dose of 10–100 viral particles. HAV is excreted in feces, and infection mainly occurs through the fecal-oral route either by direct contact with an HAV-infected person or by ingestion of contaminated drinking water and food, which account for 2–7 % of the total disease burden. Nevertheless, this figure is probably an underestimate because a considerable proportion of cases (~68 %) remain uncharacterized. Secondary spread of HAV after primary introduction by, for example, food-related contamination is common and often results in larger prolonged outbreaks (Hollinger and Emerson 2007).

Occasionally, HAV is also acquired through blood transfusions (Noble et al. 1984) or sexual contact (anal-oral). Men who have sex with men (MSM) are a known group at risk for hepatitis A. Periodic hepatitis A outbreaks among MSM have been reported since early 1990s in several developed countries (Cotter et al. 2003; Tortajada et al. 2012).

Via sewage discharge, HAV can contaminate food crops, natural watercourses, and soil. Therefore, food and drinking water are considered major vehicles of HAV transmission to humans.

Foodborne outbreaks of HAV are mainly associated to three food categories (Table 1.1). *Bivalve molluscs* are filter feeders that ingest and pass out large quantities of water to screen and consume tiny food particles. As part of this process, they can accumulate HAV if the water has been contaminated with human sewage. Shellfish are also at risk because they often are eaten raw, like oysters; improperly cooked, like most of other molluscan shellfish; or steamed for only a few minutes.

G. Sánchez, *Hepatitis A Virus in Food: Detection and Inactivation Methods*,
SpringerBriefs in Food, Health, and Nutrition, DOI 10.1007/978-1-4614-7104-2_1,
© Gloria Sánchez 2013

**Table 1.1** Selected foodborne outbreaks of hepatitis A

| Year | Location | Implicated food | Origin of raw material | Cases (n) | Comments | References |
|---|---|---|---|---|---|---|
| 2009–2010 | Australia, The Netherlands, France | Sun-dried tomatoes | Turkey | 308 | | (Gallot et al. 2011; Petrignani et al. 2010) |
| 2008 | Austria | Deli food | | 21 | Food handler contamination | (Schmid et al. 2009) |
| 2008 | Spain | Coquina clams | Peru | 100 | | (Pinto et al. 2009) |
| 2008 | China | Bottled water | China | 269 | Suspected contamination of the water source with runoff from melting of heavy snowfall | http://www.china.org.cn/government/local_governments/2008-04/24/content_15007889.htm |
| 2007 | France | Oysters | | 111 | | (Shieh et al. 2007) |
| 2005 | USA | Oysters | USA | 39 | | (Robesyn et al. 2009) |
| 2004 | Belgium | Raw beef | | 269 | Food handler contamination | (Frank et al. 2007) |
| 2004 | Egypt | Orange juice | | 351 | Poor hygiene during processing | (Wheeler et al. 2005) |
| 2003 | USA | Green onions | Mexico | 601, 3 deaths | | (Calder et al. 2003) |
| 2002 | New Zealand | Blueberries | | 39 | | FDA |
| 2001 | USA | Frozen raspberries | | | | (Fiore 2004) |
| 2001 | USA | Sandwiches | | 43 | | (Furuta et al. 2003) |
| 2001 | Japan | Clams | China | 51 | Clams were steamed with red pepper | Recall, 2000 (FDA) |
| 2000 | USA | | Frozen sliced strawberries | 7 | Berries were part of ice cream | (Bosch et al. 2001) |
| 1999 | Spain | Coquina clams | Peru | 184 | | (Dentinger et al. 2001) |
| 1998 | Ohio (USA) Mexico | Green onions | California (USA) | 43 | | |

| Year | Location | Food | Origin | Cases | Comments | Reference |
|---|---|---|---|---|---|---|
| 1997 | USA (multiple states) | Frozen strawberries | Mexico | 242 | Possible contamination during harvesting. Hand washing in field was limited. Stems were removed with fingernails | (Hutin et al. 1999) |
| 1994 | USA Georgia | Glazed donuts | | 79 | Food handler contamination | (Weltman et al. 1996) |
| 1990 | Montana | Frozen strawberries | California | 57 | Frozen strawberries used to make dessert. Suspected contamination by infected picker(s). Fruits were washed in 3 ppm chlorine before slicing and freezing | (Niu et al. 1992) |
| 1988 | Kentucky (USA) | Lettuce | Kentucky (USA) | 202 | | (Rosenblum et al. 1990) |
| 1988 | Scotland | Frozen raspberries | Scotland | 5 | Raspberries from a small farm were frozen at home. Several pickers at the farm had symptoms of hepatitis A | (Ramsay and Upton 1989) |
| 1988 | Shanghai (China) | Clams | China | 300,000 | | (Halliday et al. 1991) |
| 1983 | Malaysia | Cockles | | 322 | Raw and partially cooked cockles | (Goh et al. 1984) |
| 1983 | Scotland | Frozen raspberries | Scotland | 24 | Suspected raspberry mousse prepared from frozen raspberries. Suggested contamination by infected picker(s) | (Reid and Robinson 1987) |
| 1979 | UK | Mussels | Ireland | 41 | | (Bostock et al. 1979) |
| 1973 | USA | Oysters | USA | 265 | | (Mackowiak et al. 1976) |

*Berries and vegetables* (fresh produce) may be contaminated by irrigation water or by virus-infected individuals (e.g., seasonal workers picking berries) (Van Boxstael et al. 2013). An illustrative example is the outbreak associated with the consumption of green onions that were traced back to two farms in Mexico (CDC 2003; Dentinger et al. 2001; Milazzo and Vale 2005; Wang and Moran 2004; Wheeler et al. 2005).

*Ready-to-eat meals* can become contaminated during preparation through contact with fecally contaminated hands and surfaces. Several reported outbreaks have been associated with a immunoglobulin M–positive food handler who contaminated one or more food items, for instance donuts (Weltman et al. 1996), deli food (Schmid et al. 2009), or raw beef (Robesyn et al. 2009).

In addition to the above-mentioned food items, recent HAV outbreaks associated with imported sun-dried tomatoes (Petrignani et al. 2010) or the presence of HAV RNA on dates (Boxman et al. 2011) highlight the importance of considering other food items during outbreak investigations. Moreover, this type of outbreak warrant from foods exported from HAV endemic countries that are consumed untreated.

So far there is not much information regarding the level of HAV contamination in food products. Costafreda et al. (2006) reported between $7.5 \times 10^3$ and $7.3 \times 10^5$ HAV genomes per gram of digestive tissues, or around $1 \times 10^3$ to $1 \times 10^5$ per gram of clam. Benabbes et al. (2012) recently detected HAV in two Moroccan shellfish samples, with concentrations of around 100 RNA copies per gram of digestive tissue.

In Mexican vegetables, cilantro, parsley, and green onions, HAV levels ranged from $2.4 \times 10^3$ to $2.8 \times 10^2$ RNA copies per gram (Felix-Valenzuela et al. 2012). These viral loads are by far greater than the estimated infective dose for HAV, that is, 10–100 viral particles (Yezli and Otter 2011).

The importance of foodborne diseases caused by HAV is increasingly being recognized, and the World Health Organization has found that there is an upward trend in their incidence. This recognition is reflected by the attention that national and international organizations give to considering the control of foodborne viral infection in the report of the Advisory Committee on the Microbiological Safety of Food (Advisory Committee on the Microbiological Safety of Food 1998), the recent proposed guidelines for the application of food hygiene to the control of viruses for Codex Alimentarius (CX/FH/10/42/5), the scientific opinion of the European Food Safety Authority (EFSA) (Hazards EPoB 2011), and the expert advice on foodborne viruses for Codex Alimentarius (www.who.int/foodsafety/publications/micro/mra13/en/index.html). This latter document concluded, among other considerations, that prevention and control measures should be considered for HAV in bivalve molluscan shellfish, fresh produce, or prepared foods.

## 1.1  Classification

HAV, the prototype of the *Hepatovirus* genus within the Picornaviridae family (Fauquet et al. 2005), is composed of an icosahedral capsid that contains a positive sense, single-stranded RNA genome of approximately 7.5 kb. Its RNA genome bears different distinct regions: the 5′ and 3′ noncoding regions; the P1 region, which encodes the structural proteins VP1, VP2, VP3, and a putative VP4; and the P2 and P3 regions, encoding nonstructural proteins associated with replication.

Some degree of nucleotide sequence heterogeneity of the P1 genomic region has been observed among independent HAV isolates from different regions of the world (Costa-Mattioli et al. 2001; Lemon and Binn 1983; Robertson et al. 1992). However, this variability at the nucleotide level is not reflected in an equivalent degree of variation at the amino acid level (Hollinger and Emerson 2007; Sánchez et al. 2003b).

The high degree of conservation of the amino acid sequences of the capsid proteins of HAV entails a low antigenic diversity; therefore, only a single serotype of human HAV has been recognized (Hollinger and Emerson 2007). Moreover, Sánchez et al. (2003a) have described that, despite antigenic conservation, HAV replicates as a complex distribution of mutants, a feature of viral quasispecies.

Although only a single HAV serotype has been described, several genotypes can be differentiated from molecular methods based on the putative VP1/2A junction (Nainan et al. 2006). Initially, seven genotypes were identified: genotypes I, II, III, and VII, associated with human infections, and genotypes IV, V, and VI in simians. However, recent publications have reclassified genotype VII as a subgenotype of genotype II (Costa-Mattioli et al. 2002a; Lu et al. 2004). Genotypes I and III are the most prevalent genotypes isolated from humans. Subgenotype IA seems to be responsible for the majority of hepatitis A cases worldwide, whereas subgenotype IB viruses have been found mainly in the Mediterranean region (Nainan et al. 2006; Pinto et al. 2007), although they may be reported elsewhere too (Sánchez et al. 2002). A new subgenotype, IC, recently has been proposed (Pérez-Sautu et al. 2011). This subgenotype most probably is derived, by quasiespecies dynamics, from subgenotype IA, most common in South America.

## 1.2   Features of Hepatitis A Infection

Hepatitis A is a self-limited disease that results in fulminant hepatitis and death in only a small proportion of patients. The course of hepatitis A may be extremely variable. However, it is a significant cause of morbidity and socioeconomic losses in many parts of the world. Symptoms develop gradually and include loss of appetite, fever, headache, nausea, and vomiting, followed by jaundice 1–2 weeks later, with no associated chronic illness. The illness lasts from a few weeks to several months and is typically more severe in adults than in children, in whom it is asymptomatic. HAV infection has a long incubation period of 2–7 weeks, and individuals can excrete high numbers of virus particles ($10^6$–$10^8$ particles/g of feces) during infection. The shedding of particles can start in the last 2 weeks of the incubation period and in some individuals can continue for up to 5 months after infection (Hollinger and Emerson 2007).

Death may occur, particularly in the elderly (2.1 %). The fatality rate in HAV infections is lower than 0.1 %, although recent HAV outbreaks tend to be more severe. For example, the outbreak in 2003 associated with the ingestion of contaminated green onions caused three deaths among 601 cases (Wheeler et al. 2005).

## 1.3  Epidemiology

The incidence of HAV infection varies considerably between regions of the world, with the highest rate in developing countries where sewage treatment and hygiene practices are poor. In much of the developing world, where HAV infection is endemic, the majority of individuals are infected during early childhood, and virtually all adults are immune. In developed countries, however, HAV infections have become less common as a result of increased living standards. In these low to moderately endemic countries, only a few people are infected during early childhood, and the majority of adults remain susceptible to infection, which may lead to the occurrence of hepatitis A outbreaks (Table 1.1).

Geographic areas can be characterized by high, intermediate, or low levels of endemicity patterns of HAV infection (http://www.who.int/csr/disease/hepatitis/whocdscsredc2007/en/index4.html). The levels of endemicity correlate with the hygienic and sanitary conditions of each geographic area.

High endemicity is reported in developing countries with poor sanitary and hygienic conditions (parts of Africa, Asia, and Central and South America). In these countries infection is usually acquired during early childhood as an asymptomatic or mild infection. Reported disease rates in these areas are therefore low, and outbreaks of disease are rare. Reported disease incidence may reach 150 per 100,000 per year.

Intermediate endemicity is reported in developing countries, countries with transitional economies, and some regions of industrialized countries where sanitary conditions are variable (Southern and Eastern Europe and some regions in the Middle East). In these countries children escape infection during early childhood; however, these improved economic and sanitary conditions may lead to a higher disease incidence because infections occur in older age groups and reported rates of clinically evident hepatitis A are higher.

Low endemicity occurs in developed countries (Northern and Western Europe, Japan, Australia, New Zealand, USA, Canada) with good sanitary and hygienic conditions, and infection rates are generally low. In countries with very low HAV infection rates, disease may occur among specific risk groups such as travelers or MSM. Another fact to bear in mind is that the number of reported cases of HAV infection has declined substantially in countries with effective programs of immunization with a HAV vaccine.

## 1.4  Regulations and Recommendations

No specific legislation including microbiological criteria for viruses in foods exists in the European Union or in the USA.

Current standards for the evaluation of the sanitary quality of shellfish rely entirely on bacterial indicators of fecal contamination, either fecal coliforms or *Escherichia coli*. It has been well documented that such indicators are not correlated with the presence of viral pathogens, and HAV has been detected in shellfish from areas classified as suitable for commercial exploitation according fecal coliform criteria (Abad et al. 1997; Prato et al. 2013; Sánchez et al. 2002). Therefore, many ongoing discussions have proposed that sanitary controls of shellfish include viral parameters to guarantee safety for human consumption.

EFSA currently recommends (Hazards EPoB 2011) that European Union environmental legislation considers specific protection against fecal pollution in bivalve mollusc production areas. Control measures need to focus on avoiding fecal contamination in these areas as much as possible. Sanitary surveys would provide the necessary knowledge base. Preventive approaches could include introducing prohibition zones in the proximity of sewage discharges, using more stringent *E. coli* standards for class B classification areas, and using pollution alert procedures.

As for shellfish, no microbial criteria exists for viruses in fresh produce. The requirements of food business operators producing or harvesting plant products are general in nature and leave room for subjective interpretation, for example, to use potable or clean water whenever necessary.

The EFSA Panel on Biological Hazards (Hazards EPoB 2011) recently has concluded that controlling viruses in fresh produce clearly needs a food-chain approach, taking into account all aspects from primary production to consumption. This includes consideration of the inputs to primary production, which include the farm environment (soil, wildlife, etc.), irrigation water source, and manure. In addition, the workers (growers, pickers) and transport (open transportation may provide contamination opportunities) from the field to the packing and processing plants are a consideration at this stage. All of these stages represent potential sources of contamination, and their relevance to the particular commodity of concern may need to be assessed. At later stages, for example, during packing, the possibilities for contamination from handling needs to be taken into account.

Guidelines on the application of general principles of food hygiene to the control of viruses in food recently have been elaborated by the Codex Committee on Food Hygiene (CX/FH 10/42/5). The primary purpose of these guidelines is to minimize the risk of illness arising from the presence of human enteric viruses in foods, including specifically HAV. These guidelines focus specifically on controlling viruses in shellfish and fresh produce and include specific guidance on food handling.

The European Commission recently has issued emergency measures regarding virus testing of frozen strawberries originating in China. These measures came into place January 1, 2013, and require testing for both the norovirus and HAV.

## 1.5 Vaccination

Inactivated HAV vaccines have been available since the early 1990s and provide long-lasting immunity against hepatitis A infection. The immunity is largely related to the induction of high titers of specific antibodies. These vaccines have a high efficacy, because of the existence of a single serotype of HAV. These vaccines consist of viruses that are grown to high titers in cell culture, purified, inactivated with formalin, and adsorbed to an aluminum hydroxide adjuvant. They have quite a high economic cost, and therefore many discrepancies about their universal use in massive vaccination campaigns already exist. Countries with previous high or intermediate endemicity of HAV have performed studies of the beneficial impact of child vaccination on the overall incidence of hepatitis A, concluding that HAV immunization is economically and medically justified (Dagan et al. 2005; Domínguez et al. 2008; Wasley et al. 2005). In contrast, other countries in a similar situation do not currently recommend the implementation of such a measure in terms of the cost and related benefits (Franco and Vitiello 2003). In this context it is quite evident that highly endemic countries that usually have low economic incomes cannot afford the hepatitis A vaccination.

As a general rule, in low and intermediate endemic regions, where paradoxically the severity of the disease is high, vaccination against hepatitis A should be recommended for those in high-risk groups, including travelers to highly endemic areas, MSM, drug users, and patients receiving blood products. In addition, the inclusion of hepatitis A vaccines in mass vaccination programs in countries receiving large numbers of immigrants from endemic countries is particularly advisable.

Nevertheless, the quasiespecies replication pattern of HAV (Sánchez et al. 2003b) could lead to the selection of new antigenic variants escaping immune protection in populations with continued exposure to the virus (Aragonès et al. 2008). Hence, in these conditions, mass vaccination programs in highly endemic areas are controversial.

# Chapter 2
# Analytical Methods for Detecting the Hepatitis A Virus in Food

Because of the increasing number of hepatitis A virus (HAV) outbreaks associated with food products (Table 1.1), it has become even more important to have reliable and widely applicable techniques for detecting and quantifying HAV in food samples (reviewed by Bosch et al. 2011; Sánchez et al. 2007). Moreover, since the infectious dose of HAV is very low (10–100 viral particles) (Yezli and Otter 2011), sensitive methods are therefore needed when screening food products for the presence of HAV.

HAV detection is based on two main principles: the detection of infectious viruses by propagation in cell culture or the detection of viral genomes by molecular amplification techniques such as reverse transcriptase polymerase chain reaction (RT-PCR) or real-time quantitative PCR (RT-qPCR). HAV detection by cell culture is mainly based on the formation of cytopathic effects, followed by quantification of the viruses by plaque assay, the most probable number or tissue culture infectious dose 50. Although cell culture propagation of wild-type strains of HAV may be possible, the procedure is complex and tedious because it requires virus adaptation before its effective growth. The use of a cell line that allows the growth of a wild-type HAV isolate from stool has been reported, although its validity for broad-spectrum isolation of HAV is not yet demonstrated (Konduru and Kaplan 2006). In conclusion, until issues are resolved regarding assay complexity, cost-effectiveness, and validity for the detection of a broad spectrum of HAV isolates, infectivity is not yet a useful method for detecting HAV in food, except on the "challenges test," that is, evaluating different food manufacturing processes by inoculating food with the strain adapted by cell culture.

Current methods for the detection of HAV in food products require multiple steps: virus extraction from food; purification of the RNA (in most of the procedures a concentration step is needed); and last, molecular detection. Nucleic acid amplification techniques are currently the most widely used methods for detection and quantification of HAV in food. Among them, real-time RT-PCR assays have revolutionized nucleic acid detection by the high speed, sensitivity, reproducibility, and minimization of cross-contamination.

G. Sánchez, *Hepatitis A Virus in Food: Detection and Inactivation Methods*,
SpringerBriefs in Food, Health, and Nutrition, DOI 10.1007/978-1-4614-7104-2_2,
© Gloria Sánchez 2013

Many of the published methods or methods under development for the detection of HAV in foods use a 10- to 100-g sample size (reviews by Bosch et al. 2011 and Croci et al. 2008). These methods are generally diverse and not standardized; however, several organizations call for the development of standardized methodologies validated for their use in selected food matrices. For instance in Europe, the CEN/TC25/WG6/TAG4 working group has been entrusted by the European Committee for Standardization to establish a horizontal method for detecting the norovirus and hepatitis A virus in different types of foods and bottled water. This group has been working extensively on the procedure, and right now the analysis of food matrices for the detection of HAV is well established to the extent that European Standards have assigned an International Organization of Standardization (ISO) number and the standard is under development (ISO/ISO/PRF TS 15216-1):

- Microbiology of food and animal feeding stuffs—Horizontal method for detection of hepatitis A virus and norovirus in food using real-time RT-PCR—Part 1: Method for quantitative determination
- Microbiology of food and animal feeding stuffs—Horizontal method for detection of hepatitis A virus and norovirus in food using real-time RT-PCR—Part 2: Method for qualitative detection

Validation studies are currently underway for each of the process stages to confirm the standard.

In this standard is stated that because many food matrices contain substances that are inhibitory to RT-qPCR, it is necessary to use a virus/RNA extraction method that produces highly clean RNA preparations that are fit-for-purpose. Therefore, for soft fruit and salad vegetables the proposed method for HAV extraction is by elution with agitation followed by precipitation with polyethylene glycol (PEG)/sodium chloride. For bottled water, adsorption/elution using positively charged membranes followed by concentration by ultrafiltration is used; for bivalve molluscan shellfish, viruses are extracted from the tissues of the digestive glands using treatment with a proteinase K solution. All matrices share a common RNA extraction procedure based on virus capsid disruption with chaotropic reagents followed by adsorption of RNA to silica particles. Real-time RT-PCR is the method of choice for HAV detection and quantification. Because of the complexity of the method this standard includes a comprehensive suite of controls.

Other institutions such as Health Canada have issued a standard method for HAV detection in bottled water, green onions, and strawberries, which is published online in Health Canada's Compendium of Analytical Methods (http://www.hc-sc.gc.ca/fn-an/res-rech/analy-meth/microbio/volume5/index-eng.php).

The following sections address the issues of how to release HAV from each type of food matrix, nucleic acid extraction methods, molecular detection techniques, and quality controls.

## 2.1 HAV Extraction from Food

HAV extraction from food matrices can be defined as the separation and concentration of virus particles from the food matrix. The type of food, its composition, and the route of contamination determines how to release HAV from food before nucleic acid extraction (Bosch et al. 2011), and therefore many different procedures have been published (reviewed by Stals et al. 2012). All these methods can be grouped by food matrix.

### 2.1.1 Shellfish

Shellfish are filter feeders that can accumulate HAV in their tissues if the water has been contaminated with human sewage. Molecular HAV detection in shellfish samples involves problems with inhibitors and low virus concentrations. Nevertheless, several efficient procedures are currently available for HAV detection in molluscan shellfish, and some of them already have been applied during outbreak investigations (Pinto et al. 2009; Sánchez et al. 2002).

Table 2.1 lists different procedures for the processing of shellfish samples before the specific HAV detection by molecular procedures. Since viruses are most likely to be located in the digestive tract of shellfish (Abad et al. 1997; Romalde et al. 1994), most of these methodologies include the dissection of stomach and digestive diverticula, as initially proposed by Atmar et al. (1995). Testing the stomach and digestive tract for HAV presented several advantages in comparison with testing whole shellfish: increased test sensitivity, decrease in the sample-associated interference with RT-qPCR, and, in some instances, a less time-consuming procedure.

Alkaline elution combined with the PEG precipitation approach has resulted in varying detection limits when detecting HAV in either whole shellfish or shellfish digestive tissue (Table 2.1). Chloroform-butanol extraction, Cat-Floc elution, and PEG precipitation have been applied for confirmation of clams as the source of an HAV shellfish outbreak (Bosch et al. 2001; Costafreda et al. 2006; Sánchez et al. 2002) and for the detection of HAV in Moroccan cockles and clams (Benabbes et al. 2012).

A recent methodology combining proteinase K and heat treatment at 65 °C has been selected by the CEN/TC275/WG6/TAG4 working group for the extraction of HAV from shellfish digestive tissues (Lees 2010). This procedure is included in the draft of the ISO/TS5216 standard and has been used for confirmation of clams as the source of an HAV shellfish outbreak (Pinto et al. 2009).

After virus extraction, a variety of subsequent nucleic acid extraction and purification protocols may be employed (Table 2.1), and they are discussed in the following sections.

**Table 2.1** Methods for HAV extraction from shellfish samples

| Procedure | | Shellfish type | | |
| Elution | Concentration | (weight) | Detection limit | References |
| --- | --- | --- | --- | --- |
| Chloroform-butanol/cat-floc elution | PEG precipitation | Oysters and hard-shell clams (1.5 g of digestive tract) | ND | (Atmar et al. 1995) |
| Fluorocarbon extraction | PEG precipitation | Oysters (50 g) | 10 PFU | (Jaykus et al. 1995) |
| pH 9.5 glycine buffer, precipitation, tri-reagent, and purification of viral poly(A) | PEG precipitation | Hard-shell clams and Eastern oysters (25 g) | 0.015 PFU | (Kingsley and Richards 2001) |
| Alkaline buffer | PEG precipitation | Mussels (2 g of digestive tract) | 10 RT-PCRU | (Sincero et al. 2006) |
| Glycine buffer and chloroform extraction | Ultracentrifugation | Oysters (25 g) | 5 $TCID_{50}$/g | (Casas et al. 2007) |
| Glycine buffer | Cationically charged magnetic particles with Pathatrix | Oysters (5 g) | $2 \times 10^1$ RT-PCRU | (Papafragkou et al. 2008) |

*ND* not determined, *TCID$_{50}$* median tissue culture infectious dose, *PEG* polyethylene glycol, *PFU* plaque forming unit, *RT-PCRU* reverse transcriptase polymerase chain reaction units

## 2.1.2  Vegetables and Soft Fruits

Despite the great number of published methods for HAV extraction and concentration from produce (Croci et al. 2008; Fino and Kniel 2008a), only a small number of approaches have been validated extensively (Butot et al. 2007b), used to confirm viral contamination of produce items associated with HAV outbreaks (Calder et al. 2003; Papafragkou et al. 2008), or used for epidemiological investigations (Felix-Valenzuela et al. 2012; Kokkinos et al. 2012).

Elution concentration protocols are particularly pertinent to fruits and vegetables that are prone to contamination through sewage-contaminated surface water or infected food handlers during harvesting, packaging, or food preparation, where the viruses are likely to be on the surface of the food. These protocols are based on washing out HAV particles from the surface of the food using an appropriate buffer, usually an alkaline with a pH between 9 and 10.5, followed by concentration of the eluted viruses. The concentration methods vary widely, including PEG precipitation, ultrafiltration, ultracentrifugation, or immunoconcentration (Table 2.2).

**Table 2.2**  Methods for HAV extraction from soft fruits and vegetables

| Procedure | Food matrix (weight) | Detection limit | References |
|---|---|---|---|
| Alkaline buffer + PEG | Green onions (25 g) | 1 $TCID_{50}$ | (Guevremont et al. 2006) |
| | Lettuce (6 g) | 10 PFU | (Sair et al. 2002) |
| | Vegetables (100 g) | 50 $TCID_{50}$ | (Dubois et al. 2002) |
| | Strawberries and tomato sauce (30 g) | 14–33 PFU | (Love et al. 2008) |
| | Sun-dried tomatoes (25 g) | 1 PFU | (Martin-Latil et al. 2012) |
| Alkaline buffer + ultrafiltration | Various berries and vegetables (15 g) | 1.2 $TCID_{50}$ | (Butot et al. 2007b) |
| Alkaline buffer + ultracentrifugation | Sun-dried tomatoes (25 g) | 1 PFU | (Martin-Latil et al. 2012) |
| Alkaline buffer + cationic separation | Lettuce, strawberries, and green onions (25 g) | $10^2$ PFU | (Papafragkou et al. 2008) |
| Alkaline buffer + cationic separation | Lettuce, raspberries, strawberries, and green onions (50 g) | $10^2$ RT-PCRU | (Morales-Rayas et al. 2010) |
| Alkaline buffer + filtration | Lettuce (15 g) | ND | (Hyeon et al. 2011) |
| PBS elution + immunomagnetic separation | Green onions and strawberries (25 g) | 0.5 PFU | (Shan et al. 2005) |
| PBS + immunomagnetic separation + filtration | Lettuce and strawberries | 10 PFU | (Bidawid et al. 2000c) |
| PBS + centrifugation | Strawberries and lettuce (10 g) | ND | (Paula et al. 2010) |
| PBS elution + filtration | Vegetables (25 g) | 50 PFU | (Dubois et al. 2006) |
| PBS 2 % serum elution | Spinach | ND | (Shieh et al. 2009) |
| Neutral buffer + ultracentrifugation | Raspberries and strawberries (60 g) | $10^3$ RT-PCRU | (Rzezutka et al. 2006) |
| Neutral buffer + PEG | Various vegetables | 6.6 $TCID_{50}$ | (Sánchez et al. 2012b) |
| Guanidine buffer + polyvinylpyrrolidone | Spinach and tomatoes (25 g) | $3 \times 10^3$ PFU | (Hida et al. 2013) |

*ND* not determined, *$TCID_{50}$* median tissue culture infectious dose, *PBS* phosphate-buffered saline, *PEG* polyethylene glycol, *PFU* plaque-forming unit, *RT-PCRU* reverse transcriptase polymerase chain reaction units

Some studies have shown that ultrafiltration is more efficient than PEG precipitation for the recovery of HAV in several food items (Butot et al. 2007b; Lee et al. 2012). Moreover, Hyeon et al. (2011) observed that the recovery rate of filtration was statistically higher than PEG precipitation (47.3 % vs. 24.9 %) on lettuce samples. However, elution with an alkaline buffer and PEG concentration procedure, which is cheaper, has been selected by the CEN/TC275/WG6/TAG4 working group as the standard method when extracting HAV from produce and soft fruits.

In general, performance of these procedures has been performed by artificially contaminating food on the surface. Nevertheless, a novel mechanism of HAV contamination of green onions and spinach has been proposed. Chancellor et al. (2006) found HAV particles trapped inside growing green onions (the particles had been taken up intracellularly through the roots), even though survival of the virus was not evaluated. These findings have been confirmed recently by Hirneisen and Kniel (2012), who found that HAV or pressure-treated HAV were internalized up to 4 log RT-qPCR units. This mechanism warrants further examination, and if confirmed it will definitely change future approaches for the detection of viruses in vegetables.

When applying these approaches, the influence of individual food matrices on the overall efficiency of HAV detection is likely to be significant. HAV recovery from berry samples was consistently low compared with that from vegetables (Butot et al. 2007b; Dubois et al. 2002). This has been attributed to inactivation of the virus on the surface of the berry by a pH drop (Konowalchuk and Speirs 1975). The efficiency of detecting HAV on the surface of blueberries was higher than that for other berries, probably because it is easier to release viruses from smooth surfaces, as on blueberries, and because blueberries have a relatively thick and waxy skin, which limits the exudation of antiviral substances (Butot et al. 2007b).

The increase in mass production and distribution of food products has led to an increase in the number of multinational outbreaks. A wide range of pathogens has been associated with outbreaks related to vegetable products. Among them, the most common agent causing fresh produce–related outbreaks is are human noroviruses  (40 %), followed by *Salmonella* (18 %), *Escherichia coli* O157:H7 (8 %), *Clostridium* (6 %), HAV (4 %), and *Shigella* (4 %) (Doyle and Erickson 2008). Therefore, some recent studies have been devoted to developing methodologies to concentrate enteric viruses, including HAV, and foodborne bacterial pathogens. Brassard et al. (2011) have developed a method for the detection of bacteria and viruses in spinach based on the use of an electropositive charged membrane and a second concentration step using Amicon centrifugal units for viral detection. Using this procedure, the HAV detection limit was established as 10 plaque-forming unit (PFU)/g. Sánchez et al. (2012) recently evaluated a procedure for the simultaneous detection of several foodborne pathogens in fresh-cut vegetables by using a Pulsifier to elute the pathogens from the surface of the vegetable with buffered peptone water and concentration by PEG precipitation. The average recoveries from artificially inoculated parsley, spinach, and salad were approximately 20.7 % for HAV, whereas the detection limit was 6.6 tissue culture infectious dose 50. These procedures are useful during the epidemiological investigation of foodborne outbreaks and for routine surveillance of foodborne pathogens in vegetables.

## 2.1.3  Ready-to-Eat Foods

Ready-to-eat foods cover a wide range of food products, and as for the rest of the food matrices, several strategies for HAV detection have been applied. For instance,

HAV was artificially inoculated on hamburger samples (50 g) and processed by the sequential steps of homogenization, filtration, Freon extraction, and PEG precipitation (Leggitt and Jaykus 2000). In the study, HAV recovery ranged from 2 % to 4 % for HAV and the detection limit was established as $10^3$ PFU/50 g of food sample. Papafragkou et al. (2008) used cationically charged magnetic particles with an automated capture system (Pathatrix) to concentrate HAV from 25-g samples of artificially contaminated deli turkey and cake with frosting, reporting a detection limit of $2 \times 10^0$ RT-PCR units and $2 \times 10^2$ RT-PCR units, respectively.

### 2.1.4   Bottled Water

Although there is no epidemiological evidence that bottled water serves as a vehicle for viral diseases, some doubts were raised concerning its safety because of the reported finding of norovirus sequences in 33 % of commercially available bottled water samples sold in Switzerland (Beuret et al. 2000, 2002). However, attempts by other research groups to reproduce these results were always in vain (Butot et al. 2007a; Ehlers et al. 2004; Khanna et al. 2003; Lamothe et al. 2003; Liu et al. 2007; Sánchez et al. 2005; Villar et al. 2006), even though much larger numbers of bottled water samples and more sensitive detection methods were used. Further investigations strongly suggested that the original findings were due to artifacts and systematic mistakes (Sánchez et al. 2005).

Waterborne outbreaks of hepatitis A are unusual in developed countries, and to the best of my knowledge, only one HAV outbreak has been associated with the consumption of bottled water in China (http://www.china.org.cn/government/local_governments/2008-04/24/content_15007889.htm).

Taking into account the impact of Beuret's publications, different methodologies for routine monitoring of HAV in bottled water have been described. Moreover, the CEN/TC25/WG6/TAG4 working group has considered bottled water as a food product in the ISO/TS 15216.

Most of the described methods are based on the procedure described by Gilgen et al. (1997), where water is filtered through a positively charged membrane with a 0.45-µm pore size. Then, virus particles are released from the filter and further concentrated. Several authors have included some modifications to this procedure to increase virus recovery (Table 2.3). Direct lysis of viruses from positively charged membranes was first introduced by Beuret (2003) as a method to reduce losses during rinsing, elution, flocculation, or concentration steps before RNA extraction. This approach was automated and applied later by Perelle et al. (2009) and Schultz et al. (2011). However, Butot et al. (2013) recently reported the presence of inhibitors, depending on the composition of the bottled water, when using this direct procedure.

Another issue to consider is the viral adsorption into bottles. Butot et al. (2007a) have reported HAV adsorption into PET bottles, reaching 90 % after 20 days of storage. This adsorption was independent of the presence of microbiota of bottled water.

**Table 2.3** Most commonly used methods for detecting HAV in bottled water

| Procedure | Detection limit | References |
|---|---|---|
| +ve-charged membrane, ultrafiltration, and RNA extraction | ND | (Gilgen et al. 1997) |
| +ve-charged membrane and RNA extraction by easyMAG | 1 PFU/1.5 L; 211 $TCID_{50}$/1.5 L | (Schultz et al. 2011) |
| CIM columns, ultracentrifugation, and QIAamp Viral RNA kit | 10 $TCID_{50}$/L; 45 $TCID_{50}$/L | (Kovac et al. 2009; Schultz et al. 2011) |
| +ve-charged membrane, ultracentrifugation, and QIAamp Viral RNA kit | 361 $TCID_{50}$/L | (Di Pasquale et al. 2010a; Schultz et al. 2011) |
| +ve-charged membrane, sonication, ultrafiltration, and QIAamp Viral RNA kit | 0.2–20 $TCID_{50}$/L | (Butot et al. 2013) |

+ve *positively*, *ND* not determined, $TCID_{50}$ median tissue culture infectious dose, *PFU* plaque-forming unit

This phenomenon is important from a method standardization point of view because it makes it more difficult to prepare stable reference samples that can be used for collaborative trials. However, the authors have shown that HAV can also be desorbed from the bottle wall by rinsing the bottles with a small amount of elution buffer. This washing step has been considered within the ISO/TS 15216 standard.

## 2.2  Nucleic Acid Extraction and Purification

After HAV extraction from food, a variety of subsequent nucleic acid extraction and purification protocols may be employed. In the past decade a great number of protocols using commercially available kits have been published. A wide variety of commercial kits has been applied for nucleic acid purification, offering reliability and reproducibility, and they are quite easy to use. Most of these kits are based on the method described by Boom et al. (1990), which consists of guanidinium lysis then capture of nucleic acids on a column or bead of silica.

Numerous published studies have compared different nucleic acid extraction methods for HAV on various food matrices. Ribao et al. (2004) compared several nucleic acid extraction kits, showing that the total quick RNA cells and tissue kit from Talent was the most suitable for the detection of HAV in mussel tissues. On strawberries, the best results were achieved with Aurum Total RNA extraction kit (BioRad), obtaining a detection limit of 5–6.25 PFU/mg of tissue (Bianchi et al. 2011). A slightly lower sensitivity was rendered by the RNeasy Plant mini kit from Qiagen (10–12.5 PFU/mg of tissue), whereas the Total Quick RNA Cells and Tissues kit version mini from Talent rendered a detection limit of 50–100 PFU/mg of tissue.

In the past 10 years automated nucleic acid extraction platforms have been developed by commercial companies and have been shown to be suitable for the analysis of HAV in samples of bottled water (Perelle et al. 2009).

## 2.3   HAV Detection in Food by Molecular Techniques

Molecular detection procedures are used widely in the field of food virology and are continuously evolving. For instance, Sánchez et al. (2007) summarized published RT-qPCR methods for HAV detection in food, and since then several new methods have become available (Table 2.4).

Although nucleic acid sequence-based amplification (Jean et al. 2001) and loop-mediated isothermal amplification techniques (Yoneyama et al. 2007) have been reported as highly sensitive and specific, RT-qPCR remains the current gold

**Table 2.4**  Molecular methods most commonly used for HAV detection

| Target region | Molecular detection methods | Sensitivity | References |
|---|---|---|---|
| | ***RT-qPCR*** | | |
| 5′ NCR | Molecular Beacon probe | 1 PFU/reaction | (Abd El Galil et al. 2005) |
| VP1 | SYBR Green | ND | (Brooks et al. 2005) |
| 5′ NCR | TaqMan and molecular beacon | 0.05 $TCID_{50}$/reaction 1 copy/reaction | (Costafreda et al. 2006) |
| 5′ NCR | TaqMan | 5 copies/reaction | (Costa-Mattioli et al. 2002b) |
| 5′ NCR | TaqMan | ND | (Hewitt and Greening 2004) |
| 5′ NCR | TaqMan | 40 geq/reaction 0.5 PFU/reaction | (Jothikumar et al. 2005) |
| 5′ NCR/VP4 | TaqMan | ND | (Silberstein et al. 2003) |
| 5′ NCR | TaqMan | 60 geq/mL | (Villar et al. 2006) |
| Different regions | TaqMan | 1 geq/mL | (Houde et al. 2007) |
| 5′ NRC | TaqMan | 5 $TCID_{50}$/mL | (Di Pasquale et al. 2010b) |
| Not specified | FRET probes | 5 geq/reaction 0.02 $TCID_{50}$/reaction (Sánchez et al. 2006) | LightCycler HAV quantification kit (Roche Diagnostics, GmbH, Mannheim, Germany) |
| VP1 | FRET probes | 2 geq/reaction 0.05 $TCID_{50}$/reaction | RealArt HAV LC RT PCR kit (Artus GmbH, Hamburg, Germany) (Sánchez et al. 2006) |
| | **Other assays** | | |
| 5′ NCR | Nested real-time PCR | 0.2 PFU | (Hu and Arsov 2009) |
| VP1–VP2 | NASBA | $4 \times 10^2$ PFU/mL | (Jean et al. 2001) |
| 5′ NCR | LAMP | 0.4–0.8 FFU/reaction | (Yoneyama et al. 2007) |

*ND* not determined, $TCID_{50}$ median tissue culture infectious dose, *PFU* plaque-forming unit, *FFU* focus-forming unit, *geq* genome equivalent, *RT-qPCR* real-time quantitative polymerase chain reaction, *LAMP* loop-mediated isothermal amplification, *NASBA* nucleic acid sequence-based amplification standard for HAV detection and quantification in food. Currently, several RT-qPCR assays, which have revolutionized nucleic acid detection by the high speed, sensitivity, reproducibility, and minimization of contamination, have been described for HAV detection (Bosch et al. 2011; Sánchez et al. 2007). Some of them are even commercially available, for example, the LightCycler hepatitis A virus quantification kit (Roche Diagnostics), the RealArt HAV LC RT PCR kit (Artus GmbH), and the KHAV for environmental and food samples (Ceeram).

When standardizing methodologies, it is essential that the specificity and sensitivity of RT-qPCR assays are demonstrated. All these points are interconnected and depend mostly on the target sequences for primers and probe. The selected targets must guarantee an absolute specificity and must reach equilibrium between high sensitivity, broad reactivity, and reliability of quantification.

Unlike SYBR Green, an intercalating dye that binds to all doubled-stranded DNA, probe-based RT-qPCR (e.g., TaqMan, molecular beacon, or FRET probes) uses a fluorescently labeled, target-specific probe that results in increased specificity and sensitivity. Assays based on the TaqMan chemistry (Costa-Mattioli et al. 2002b; Costafreda et al. 2006; Hewitt and Greening 2004; Jothikumar et al. 2005; Silberstein et al. 2003; Villar et al. 2006) are the methods that have been commonly used for the detection and quantification of HAV. Although some authors have used primers targeted to the VP1 capsid region (Brooks et al. 2005), primers targeting the highly conserved 5′ noncoding region are more widely preferred (Table 2.4), most of which show similar performance (approximately 1 RNA copy per reaction). Nevertheless, a good validation study of the selected primers must be carried out to avoid failing in the detection of specific genotypes, as previously described (Heitmann et al. 2005). Of all the above-mentioned RT-qPCR assays, the one suggested by Costafreda et al. (2006) is the method most extensively used for HAV detection in food (Fraisse et al. 2011; Martin-Latil et al. 2012; Sánchez et al. 2011) and is currently included in the procedure proposed by the CEN/TC/WG6/TAG4 group. This RT-qPCR assay recently has been incorporated in a one-step, duplex real-time qRT-PCR (Blaise-Boisseau et al. 2010). Likewise, new developments are moving now to combine the detection of several foodborne pathogens in a single reaction tube. For instance, Morales-Rayas et al. (2010) have designed a multiplex RT-qPCR assay for the detection of HAV and norovirus genotype I and genotype II. The detection limit of this multiplex RT-qPCR was determined as 10 HAV particles.

## 2.4  Quality Controls

Although the detection of HAV in food is mainly performed by RT-qPCR techniques, there are several limitations. These types of methodologies are prone to inhibition, favoring false-negative results and demonstrating the need for proper quality controls (D'Agostino et al. 2011).

Several methods to overcome the presence of inhibitors have been described, such as the analysis of diluted samples, smaller food sample sizes, dissection of the digestive tract of shellfish samples, addition of pectinase to berry samples, or adaptation of the RT-qPCR by, for example, the addition of Tween, bovine serum albumin, or commercial reagents (Bosch et al. 2011).

Another obstacle that influences HAV detection in foods is the low efficiency of concentration and nucleic acid extraction procedures, which influence the performance of the whole methodology.

To monitor all these issues, quality control and quality assurance measures include the use of positive and negative controls, thus tracing any false-negative or false-positive result, respectively. Most false negatives are a consequence of inefficient HAV extraction or inhibition of the RT-qPCR reaction. Most false positives result from cross-contamination (D'Agostino et al. 2011). Therefore, the most recent methods include an internal amplification control, an added external control for monitoring extraction efficiency, or both. For instance, Costafreda et al. (2006) have proposed the use of a nonpathogenic virus, the mengovirus MC0 strain, as a HAV process control. In this approach a titrated process control is added during the initial step of the procedure and is quantified by RT-qPCR at the end of the procedure, allowing an accurate estimate of the performance of the whole procedure. This approach already has been used by several authors to evaluate the efficacy of their procedures as well as for accurate quantification of HAV in naturally contaminated food samples (Benabbes et al. 2012; Butot et al. 2009; Pinto et al. 2009).

The feline calicivirus also has been proposed as a process control for HAV detection in food and water samples (Di Pasquale et al. 2010b; Mattison et al. 2009). However, the feline calicivirus has been reported to be an inappropriate surrogate for HAV in acid conditions (Butot et al. 2008; Cannon et al. 2006).

## 2.5 Assessment of HAV Infectivity

Viral infectivity is defined as the capacity of a virus to enter the host cell and exploit its resources to replicate and produce infectious viral particles, which may lead to infection and subsequent disease in the human host (Rodriguez et al. 2009). Obviously, cell culture–based methods are the soundest methodologies for the estimation of the number of infective HAV particles. However, as indicated earlier, there are no available culture models for routine monitoring of wild-type HAV strains. Therefore, cell culture–based methods have been used mainly in artificially inoculated experiments for method development or evaluation of inactivation processes.

One of the major limitations of using PCR or real-time PCR is its inability to differentiate between infectious and noninfectious viruses. This is relatively important when detecting HAV in food samples subjected to food manufacturing processes. In some instances the reduction in the number of infectious viruses did not correlate with the number of genomes detected by real-time RT-PCR (Butot et al. 2008, 2009; Hewitt and Greening 2004).

Several different approaches to assess HAV infectivity using PCR have been attempted (revised by Rodriguez et al. 2009). The detection of an intact HAV genome or specific region may indicate that the virus capsid is protecting the genome from degradation and therefore determining infectivity. For instance, Bhattacharya et al. (2004) reported that the amplification of the 5′ and 3′ noncoding regions may be suitable to discriminate between infectious and noninfectious HAV after ultraviolet and heat treatments. Moreover, it has been suggested that the 5′ noncoding region is the most easily degraded region of the genome of HAV upon exposure to chlorine and chlorine dioxide, and it has been used to discriminate between infectious and noninfectious HAV (Li et al. 2002; Sánchez et al. 2012).

Another approach is to assess damage of the capsid, which results in loss of protection of the HAV genome. In this approach, enzymatic treatment with RNase, in some instances combined with a proteinase K treatment, have been applied to differentiate successfully between an intact HAV and HAV inactivated by ultraviolet light, chlorine disinfection, and thermal treatment at 72 °C (Nuanualsuwan and Cliver 2003).

A promising new strategy to assess HAV infectivity relies on the use of nucleic acid intercalating dyes such as propidium monoazide (PMA) or ethidium monoazide as a sample treatment before the RT-qPCR. This approach recently has been applied to discriminate successfully between infectious and noninfectious HAV inactivated by heat treatment (Sánchez et al. 2012). As suggested by Parshionikar et al. (2010) for other enteric viruses, stable secondary structures of the RNA genome may facilitate covalent binding of PMA to viral RNA. This is in agreement with the successful application of PMA RT-qPCR for enterovirus (Parshionikar et al. 2010) and HAV (Sánchez et al. 2012), where the selected primer sets are targeted to the 5′ noncoding region.

It recently has been reported that PMA treatment was more effective than RNase treatment for differentiating infectious and thermally inactivated HAV (Sánchez et al. 2012). Results showed that combining 50 µM of PMA and RT-qPCR selectively quantify infectious HAV in media suspensions. Although this methodology has not been evaluated in food samples, PMA treatment before RT-qPCR detection is a promising alternative to assess HAV infectivity in food samples.

# Chapter 3
# HAV Survival and Inactivation Under Different Food Processing Conditions

In this chapter, the stability of HAV in food and its inactivation under the most common food processing conditions will be compiled.

Summarizing, HAV is unusually stable to acid treatments, demonstrating infectivity following exposure to a pH of 1.0 for 2 h. It is relatively resistant to heat, being only partially inactivated after 12 h at 60 °C. HAV is resistant to drying and remains infectious for several days to months in contaminated fresh water, seawater, wastewater, soils, oysters, and cookies. HAV can be inactivated by iodine (3 mg/L for 5 min) and sodium hypochlorite (3 mg/L at 20 °C for 5 min). HAV is not destroyed by freezing and remains stable for several years at temperatures of −20 °C to −70 °C. HAV inoculated in bottled water is still infectious after 300 days (reviewed by Baert et al. 2009; Sattar et al. 2000; Rzeutka and Cook 2004).

## 3.1 Stability of HAV in Food Products

HAV is known to tolerate environmental factors more than many other viruses, including most enteric viruses. HAV is a highly stable virus, able to persist for extended periods of time in the environment, and therefore in food products. HAV has been shown experimentally to be able to survive in several food products and can persist under normal storage conditions over the usual periods between purchase and consumption.

**Chilled storage temperatures** (2–8 °C) typically retard respiration, senescence, product browning, moisture loss, and microbial growth in minimally processed fruits and vegetables, but may contribute to the survival and transmission of HAV (Seymour and Appleton 2001). Survival of HAV on chilled foods has been studied extensively (reviewed by Baert et al. 2009). Most of the studies found that HAV remained infectious for periods exceeding the shelf-life of products. On vegetables, Croci et al. (2002) evaluated HAV survival on carrots and fennel, reporting complete inactivation of HAV by day 4 for carrots and by day 7 for fennel. Sun et al. (2012)

G. Sánchez, *Hepatitis A Virus in Food: Detection and Inactivation Methods*,
SpringerBriefs in Food, Health, and Nutrition, DOI 10.1007/978-1-4614-7104-2_3,
© Gloria Sánchez 2013

recently showed that HAV survived during more than 20 days during storage at 3–10 °C on contaminated green onions. Shieh et al. (2009) investigated the survival of HAV on fresh spinach leaves in moisture- and gas-permeable packages that were stored at 5.4 °C for up to 42 days, reporting only 1-log reduction of HAV infectivity over 4 weeks of storage.

On shellfish, HAV inoculated in commercially prepared marinated mussels showed a 1.7-log reduction of HAV infectivity after 4 weeks of storage at 4 °C (Hewitt and Greening 2004).

The stability of HAV inoculated in bottled water was studied at 4 °C (Biziagos et al. 1988). Infectious HAV was detected after 1 year of storage, with only a 0.68-log reduction. Biziagos et al. also found that HAV stability was dependent on the protein concentration.

In general, the above-mentioned studies indicated that HAV will survive on chilled food over the periods before deterioration of the specified food.

**Relative humidity.** The influence of relative humidity on HAV survival on vegetables and fruits has been investigated. Stine et al. (2005) investigated the survival of HAV on bell peppers, cantaloupe, and lettuce stored at 22 °C under high (mean, 85.7–90.3 %) and low (mean, 45.1–48.4 %) relative humidity. HAV survived significantly longer on cantaloupe than on lettuce, and high inactivation rates were reported under conditions of high humidity.

**Frozen storage.** The occurrence of HAV outbreaks due to the consumption of shellfish and berries that had been frozen several months (Bosch et al. 2001; Hutin et al. 1999; Niu et al. 1992; Pinto et al. 2009; Ramsay and Upton 1989; Reid and Robinson 1987; Sánchez et al. 2002) indicates that if food is contaminated before freezing, substantial fractions of the viruses will remain infectious during frozen storage. Moreover, some studies have reported the presence of HAV sequences on frozen shellfish. For instance, Shieh et al. (2007) were able to detect and type HAV sequences in oysters implicated in an outbreak. Those oysters were stored in the cold for 12 days and then frozen at −20 °C for 7 weeks before analysis. Sánchez et al. (2002) also detected and typed HAV from imported frozen clams that caused an outbreak in Spain affecting 184 people. All these results indicate that freezing has little or no effect on HAV infectivity in molluscan shellfish.

Butot et al. (2008) extensively investigated the survival of HAV on frozen strawberries, blueberries, raspberries, parsley, and basil, concluding that frozen storage for 3 months had no effects on HAV infectivity. Overall, these studies showed that freezing will not ensure an adequate reduction of HAV if present in foods.

**Drying.** HAV survival in dried state has been studied mostly on inanimate surfaces or fomites (Abad et al. 1994; Mbithi et al. 1991). The HAV outbreak associated with the consumption of sun-dried tomatoes indicates that if food is contaminated before drying, substantial fractions of the viruses will still remain infectious (Gallot et al. 2011; Petrignani et al. 2010).

**Modified atmosphere packaging** (MAP) is usually done to slow the respiration rate of fruits and vegetables and therefore reduce the product's metabolism and maturation. MAP is a way of extending the shelf life of fresh food products by inhibiting spoilage by bacteria and fungi. This technology substitutes the atmospheric air inside a package with a protective gas mix. Overall this type of packaging is designed to inhibit bacterial or fungal growth and therefore is not effective against HAV because it does not grow in foods. For instance, Bidawid et al. (2001) evaluated the effect of various modified atmospheres on the survival of HAV on lettuce stored at room temperature and 4 °C for up to 12 days in ambient air and under various modified atmospheres. The lettuce samples were stored in heat-sealed bags with the following percentages of gas mixtures (carbon dioxide [$CO_2$]:$N_2$): 30:70, 50:50, 70:30, and 100 % $CO_2$. Only at 70 % $CO_2$ at room temperature was a significant decline in virus survival observed. Because most commercially distributed vegetables are stored at lower $CO_2$ concentrations and at 4 °C, standard MAP conditions will not provide protection against HAV transmission.

**Acidification.** Dressings, sauces, marinades, and similar food products depend on their acidity to prevent spoilage. They may consist of naturally acidic foods, such as fruit juice or tomatoes, or they may be formulated by combining acidic foods with other foods to achieve the desired acidity. Some foods, such as vinegar and certain pickled vegetables, may develop acidity from microbial fermentation. However acidification of food is not a suitable hurdle to control HAV in foods since HAV is highly stable at an acidic pH. For instance, HAV had a high residual infectivity after 2 h of exposure to pH 1 at room temperature, remaining infectious for up to 5 h. At 38 °C and pH 1, HAV remained infectious for 90 min (Scholz et al. 1989). Therefore, acidification would not be a suitable strategy to reduce the number of viruses present on food.

## 3.2 HAV Inactivation Under Different Food Processing Technologies

Koopmans and Duizer (2004) summarized the risks of consuming products that may have become contaminated with enteric viruses. The risks were categorized as negligible, low, medium, and high, depending on whether food processing resulted in reductions in the infectivity of common foodborne viruses of at least 4, 3, 2, and 1 log units, respectively.

Several studies have investigated the efficacy of commercial processes to inactivate or eliminate HAV (Table 3.1). Most of these studies have been performed with HAV suspensions of the HM-175 strain, and little information is available about naturally contaminated food samples. Overall it is difficult to draw general conclusions from these studies because of differences in the experimental conditions and methods that were used.

This chapter looks first at the existing studies concerning HAV behavior during food processing. Results from the studies are presented and discussed and conclusions are drawn about the efficacy of different food processing technologies if raw materials are contaminated, which may allow risk to be established.

**Table 3.1** HAV inactivation by thermal treatments in food products

| Thermal treatments | Food matrix | Inactivation (log reduction) | References |
|---|---|---|---|
| **Shellfish** | | | |
| 95 °C for 1.5 min | Cockles | >4 | (Koopmans and Duizer 2004) |
| 90 °C for 10 min | Manila clams | >4 | (Cappellozza et al. 2012) |
| Boiling water for 3 min | New Zealand greenshell mussels | >4 | (Hewitt and Greening 2006) |
| Steaming for 180 s (internal temperature of 63 °C) | New Zealand greenshell mussels | 1.5 | (Hewitt and Greening 2006) |
| 90 °C for 1.5 min | Soft-shell clams | 2.7 | (Sow et al. 2011) |
| 90 °C for 3 min | Soft-shell clams | >4 | (Sow et al. 2011) |
| Marination (70 °C for 2.4 min, acetic acid) | Mussels | <2 | (Hewitt and Greening 2004) |
| 60 °C for 10 min | Mussels | 2 | (Croci et al. 1999) |
| 80 °C for 3 min | | 2 | |
| 80 °C for 10 min | | 4 | |
| Mussels hors d'oeuvre (cooking for 9 min) | Mussels | >3 | (Croci et al. 2005) |
| Mussels au gratin (opened mussels, put in a baking pan, were covered with a mixture of onions, garlic, parsley, breadcrumbs, and butter). The mussels were grilled in the oven (temperature of 250 °C for 5 min) | Mussels | >3 | (Croci et al. 2005) |
| Mussels in tomato sauce (oil, garlic, tomato sauce, and water were cooked for 15 min, then mussel bodies, oregano, and parsley were added and cooked again for 8 min at boiling temperature) | Mussels | >4 | (Croci et al. 2005) |
| Steaming | Green mussels | <2 | (Hewitt and Greening 2006) |
| Boiling water for 3 min | Green mussels | >3.5 | (Hewitt and Greening 2006) |
| **Other food products** | | | |
| High temperature–short time and immediate packing (71.7 °C, 15 s) | Milk, ice cream | <2 | (Bidawid et al. 2000d) |
| Pasteurization of solid foods (70 °C, 2 min) | Paté and other cooked meats | <2 | (Koopmans and Duizer 2004) |
| Ultrahigh temperature processing and aseptic filling (>120 °C) | Milk, dairy products | >4 | (Koopmans and Duizer 2004) |
| 85 °C for 5 min | Strawberry mashes | <2 | (Deboosere et al. 2004) |
| Blanching (95 °C, 2.5 min) | Basil, chives, mint, and parsley | >3 | (Butot et al. 2009) |
| Blanching (75 °C, 2.5 min) | Basil, chives, mint, and parsley | 2 | (Butot et al. 2009) |

**Thermal processing** is commonly used in the food industry to inactivate foodborne pathogens or spoilage microorganisms, but in general it has been designed to kill bacteria. While thermal inactivation of bacterial foodborne pathogens has been studied thoroughly in food industry, heat inactivation of HAV in food has been poorly investigated.

Inactivation of HAV by heat treatment, as shown in Table 3.1, varies with the type of food and cooking method. For instance, the presence of fats or proteins in shellfish plays a protective role (Croci et al. 1999; Millard et al. 1987).

In general, conventional pasteurization (e.g., 63 °C for 30 min or 70 °C for 2 min) seems more effective than high temperature–short time pasteurization (e.g., 71.7 °C for 15–20 s); however, HAV is unlikely to be inactivated completely during those treatments. On mussel homogenates heated at 60 °C for 10 min, HAV titers were reduced by 2 log, and infectious HAV was still detected when heat treatment was extended to 30 min (Croci et al. 1999).

The occurrence of HAV outbreaks due to the consumption of grilled, stewed, steamed, and fried shellfish indicates that regular cooking does not guarantee the complete inactivation of HAV (Koff and Sear 1967; Lees 2000). Some studies have investigated the efficacy of light cooking on HAV inactivation in shellfish. Light cooking of shellfish usually involves heating until the shell is opened, which is usually achieved at temperatures below 70 °C for approximately 47 s, which is unable to completely inactivate HAV (Abad et al. 1997). Other research studies have shown the persistence of viruses during common cooking practices of shellfish, such as steaming (Croci et al. 1999, 2005; Hewitt and Greening 2006). To conclude, complete inactivation of HAV in shellfish is achieved after steaming the shellfish to an internal temperature of 85–90 °C for 1.5 min, which may be not acceptable from an organoleptic point of view (Lees 2000).

As in shellfish, the effect of fat content has an effect on the resistance of HAV to heat in other food matrices. For instance, Bidawid et al. (2000d) evaluated HAV heat resistance in various types of milk with different fat concentrations, showing that routine pasteurization temperatures are not adequate to inactivate HAV in dairy products. Moreover, it has been shown that increasing the fat content seemed to play a protective role and can further contribute to increased stability of HAV in food products under heating conditions.

Sugar concentration also has been reported to have a big effect on HAV heat resistance in berries (Deboosere et al. 2004). Because berries and vegetables usually are consumed fresh or frozen, limited thermal inactivation data exist for HAV in this type of product. For instance, Deboosere et al. (2004) defined the heat inactivation kinetics of HAV in a fruit model system. Later, the same research group developed a model for the inactivation of HAV in red berries without supplemented sugar and with different pH values (Deboosere et al. 2010).

Butot et al. (2009) investigated the resistance of HAV to heat in freeze-dried berries: different types of freeze-dried berries were heated at 80 °C, 100 °C, or 120 °C for 20 min, showing that dry heat treatment for 20 min at 100 °C or 120 °C was capable of inactivating infectious HAV.

**Low-heat dehydration** in the 40–60 °C range continued overnight for up to 24 h has been a common means to dehydrate fruits and vegetables in households and the food industry. Effects on HAV survival after low-heat dehydration of green onions at 45 °C or 65 °C for 20 h recently have been reported (Laird et al. 2011; Sun et al. 2012). Twenty hours of heating at 47.8 °C, 55.1 °C, and 62.4 °C reduced infectious HAV in green onions by 1, 2, and 3 log, respectively. Only when the temperature reached an average of 65 °C for 20 h was a 4-log reduction of HAV infectivity reported.

**Shellfish depuration** is a commercial processing strategy that originated more than a century ago. Shellfish are placed in tanks of clean seawater and allowed to purge the contaminants over a period of several days (Richards et al. 2010). Methods for shellfish depuration have been designed to remove bacteria, but are poorly effective on HAV (Abad et al. 1997; Bosch et al. 1994; Lees 2000). Abad et al. (1997) have evaluated commercial depuration conditions used by the shellfish industry, and infectious HAV were recovered from bivalves after 96 h of immersion in a continuous flow of ozonated marine water.

Ultraviolet (UV) irradiation is the method of choice for shellfish depuration in many countries. De Abreu Corrêa et al. (2012) have shown a 3-log reduction of HAV titers in a shellfish depuration system with UV treatment. De Medici et al. (2001) evaluated the effectiveness of a closed-circuit depuration system that used both ozone and UV light for disinfecting water on mussels experimentally contaminated with HAV. The authors concluded that although HAV titers were significantly reduced, a residual amount of infectious HAV was still detected. De Abreu Corrêa et al. (2012) recently observed that after 96 h of UV treatment working in a depuration system, HAV was not detected on oyster samples. Overall, these studies indicates that shellfish depuration contributes to reduced HAV levels and hence reduced risk of infection from shellfish consumption (Bosch et al. 1994).

**Blanching** is a critical industrial food preparation process that consists of scalding vegetables in boiling or steaming water for a short time, and it helps retain the flavor, color, and texture of vegetables by stopping enzyme action. Almost all vegetables that are to be frozen must be blanched, and blanching is also recommended for vegetables before dehydration. Blanching conditions vary, and the temperatures are generally between 75 °C and 105 °C. So far, only one study has evaluated this technology on basil, chives, mint, and parsley (Butot et al. 2009). In this study, HAV titers were significantly reduced immediately following steam blanching at 95 °C for 2.5 min, and no residual infectious HAV were detected. Appreciably less inactivation of HAV was observed when the temperature of steam blanching was reduced to 75 °C, and the infectious HAV titers were reduced by an average of 2 log10 units, except for chives.

**High pressure processing (HPP)** has emerged as a promising technology because it can inactivate and inhibit microorganisms while only minimally degrading the quality of the product being treated (Rodrigo et al. 2007). Inactivation of HAV by HPP has been studied extensively using HAV suspensions (reviewed by Kovač et al. 2010), while the efficacy on food matrices is limited (Table 3.2). Treatments with at least 400 MPa are in general very efficient, but the reductions are significantly different when using different processing temperatures, times, or food matrices (Table 3.2). Moreover, organoleptic alterations need to be considered when HPP is applied to

**Table 3.2** HAV inactivation by high hydrostatic pressure and irradiation in food products

| | Treatment | Food matrix | Inactivation (log or PFU reduction) | References |
|---|---|---|---|---|
| **Hydrostatic pressure (MPa)** | | | | |
| 300 | 1 min, 9 °C | Oysters | 0.2 | (Calci et al. 2005) |
| 325 | | | 0.8 | |
| 350 | | | 1.3 | |
| 375 | | | 2.3 | |
| 400 | | | 3.2 | |
| 300 | 5 min, 18–22 °C | Mediterranean mussels | 0.1 | (Terio et al. 2010) |
| 325 | | | 0.7 | |
| 350 | | | 1.7 | |
| 375 | | | 2.5 | |
| 400 | | | 2.9 | |
| 300 | 5 min, 18–22 °C | Blue mussels | 0.8 | (Terio et al. 2010) |
| 325 | | | 1.0 | |
| 350 | | | 2.1 | |
| 375 | | | 2.7 | |
| 400 | | | 3.6 | |
| 250 | 5 min, 21 °C | Strawberry puree | 1.2 | (Kingsley et al. 2005) |
| 275 | | | 2.1 | |
| 300 | | | 3.1 | |
| 375 | | | 4.3 | |
| 250 | 5 min, 21 °C | Sliced green onions | 1.2 | (Kingsley et al. 2005) |
| 275 | | | 2.1 | |
| 300 | | | 3.1 | |
| 375 | | | 4.3 | |
| 500 | 5 min, 4 °C | Pork sausage product | 3.23 | (Sharma et al. 2008) |
| Hydrodynamic pressure | – | Pork sausage product | 1.10 | (Sharma et al. 2008) |
| **UV irradiation** (mW s/cm²) | | | | |
| 40 | | Lettuce, green onions, and strawberries | 4.3; 4.1; 1.3 | (Fino and Kniel 2008b) |
| 120 | | Lettuce, green onions, and strawberries | 4.5; 5.3; 1.8 | (Fino and Kniel 2008b) |
| 240 | | Lettuce, green onions, and strawberries | 4.6; 5.6; 2.6 | (Fino and Kniel 2008b) |
| **Gamma irradiation** (kGy) | | | | |
| 3 | | Lettuce and strawberries | 1 | (Bidawid et al. 2000b) |
| 10 | | Lettuce and strawberries | 3.3 | (Bidawid et al. 2000b) |
| 2 | | Oysters and clams | 2 | (Mallett et al. 1991) |

some food products. The mechanism of inactivation of HAV using HPP treatment has not been elucidated clearly, but results based on RNase and propidium monoazide treatments suggest that the HAV capsid remains intact following inactivation by HPP (Kingsley et al. 2002; Sánchez et al. 2012). Therefore, the mechanism of HPP inactivation for HAV is most likely due to subtle alterations of viral capsid proteins preventing attachment to the cellular receptor or the blockage of the penetration and virion-uncoating mechanisms subsequent to viral attachment (Kingsley et al. 2002).

**Vacuum freeze-drying** is considered the reference process for manufacturing high-quality dehydrated products. The production of freeze-dried products involves preliminary freezing of fresh food, followed by placing the food samples under reduced pressure with sufficient heat to sublimate ice. Vacuum freeze-drying treatment is the best method to dehydrate berries to maintain the color, flavor, and most types of antioxidants. Freeze-dried berries are commonly used by the food industry in cereals, muesli bars, chocolate products, and bakery goods. So far there has not been much research into the efficacy of freeze-drying on enteric viruses. Just one published study has reported less than 2 log reduction of HAV infectivity after freeze-drying of different types of berries and herbs (Butot et al. 2009).

### 3.2.1  Other Manufacturing Processes

**Gamma irradiation** is an alternative technology for inactivating pathogens in foods, especially produce. Nowadays, World Health Organization standards, which are based on nutritional, toxicological, and microbiological criteria, indicate that irradiation doses of $\leq 10$ kGy are considered safe. The Food and Drug Administration has approved the use of irradiation up to 4 Gy for of fresh foods to control foodborne pathogens and extend the shel-life of fresh iceberg lettuce and spinach (reviewed by Goodburn and Wallace 2013).

Studies of the effect of gamma irradiation on HAV infectivity in foods are limited. Bidawid et al. (2000b) investigated the efficacy of irradiation on HAV inactivation in strawberries and lettuce at doses ranging between 1 and 10 kGy. They observed a linear decrease in HAV titer as irradiation doses increased for both food products. At 10 kGy, more than 3 log of HAV inactivation occurred for both strawberries and lettuce. Gamma irradiation (2 kGy) applied to shellfish, oysters, and clams significantly reduced HAV levels but did not affect organoleptic qualities (Mallett et al. 1991) (Table 3.2).

*UV light* is electromagnetic radiation with wavelengths shorter than visible light that can induce damage in a broad range of microorganisms when applied as UV-C light (254 nm). Fino and Kniel (2008b) investigated the effect of UV light at various doses (40, 120, and 240 mWs/cm$^2$) on HAV inoculated on fresh produce (strawberries, green onions, and lettuce). Inactivation of HAV varied depending on the UV dose and the food product. For instance, inactivation of HAV on strawberries was the least effective, with only a 2.6-log tissue culture infectious dose 50/mL inactivation at 240 mWs/cm$^2$, compared with green onions and lettuce, where >4.5-log tissue culture infectious dose 50/mL inactivation was observed at the same UV dose. As described for other technologies, the type of surface has the greatest impact on HAV inactivation: smoother surfaces are the most easy to decontaminate. The more crevices or the presence of hairlike projections in the surface of the fruits or vegetables, the more protected are the viruses from UV light.

*Pulsed-electric field* (PEF) processing is a nonthermal method of pasteurizing liquid foodstuffs; it uses short bursts of electricity and can inactivate spoilage and pathogenic microorganisms at or near atmospheric temperature (Puértolas et al. 2012). Several authors have reported the sensitivity of a range of foodborne pathogens to PEF treatments (Haughton et al. 2012; Mukhopadhyay and Ramaswamy 2012), but so far no information is available for HAV.

**Antimicrobial packaging**. The addition of antimicrobial agents into food packaging can be used to control the food microbiota and target specific microorganisms to provide higher safety and enhance the quality of food products. As a means of preventing recontamination with pathogens and extending the shelf-life of foods, antimicrobial packaging is one of the most promising technologies in the food industry. However, few studies have confronted the task of evaluating materials with virucidal properties in real food applications. For the first time, Martínez-Abad et al. (2013) have evaluated active renewable packaging materials for virus control in vegetables using a norovirus surrogate. Further research is needed to evaluate the use of antimicrobial packaging on relevant foodborne viruses, such as HAV.

## 3.2.2 Efficacy of Washing Procedures to Eliminate/Inactivate HAV on Berries and Vegetables

The consumption of fresh-cut vegetables has increased globally because they generally are considered safe and healthy by consumers (Lynch et al. 2009). However, agricultural irrigation with wastewater that may be raw, treated, or partially diluted is a common practice worldwide and constitutes the main source of pathogen contamination. Several factors affect microbial quality and shelf-life of berries and vegetables, such as intrinsic properties of the produce (e.g., pH, water content); processing factors (e.g., washing, cutting, blanching); extrinsic factors (e.g., storage temperature, packaging); and implicit factors (e.g., microbial characteristics).

Vegetables, including various types of salads and green onions, and berries have been associated with outbreaks of hepatitis A (Calder et al. 2003; Dentinger et al. 2001; Hutin et al. 1999; Niu et al. 1992; Ramsay and Upton 1989; Reid and Robinson 1987; Rosenblum et al. 1990). Recently, a hepatitis A outbreak caused by the ingestion of contaminated green onions resulted in three deaths among a total of 601 cases (Wheeler et al. 2005). Moreover, HAV has been detected in market lettuce (Hernandez et al. 1997; Monge and Arias 1996; Pebody et al. 1998), in 1.32 %

(4/304) of salad vegetables from European countries (Kokkinos et al. 2012), and in 28.2 % of the samples from Mexico (Felix-Valenzuela et al. 2012).

Produce may be contaminated with enteric viruses during cultivation and before harvest by contact with inadequately treated sewage or water polluted by sewage. Contamination may also occur during processing, storage, distribution, or final preparation. This could happen because food is contacted by HAV-infected people, contaminated water, or fomites. In fact, experimental studies have shown that approximately 9.2 % of infectious HAV can be transferred to lettuce from the contaminated hands of the handlers (Bidawid et al. 2000a).

Although it is usually accepted that preventive measures to avoid and reduce pathogen contamination are the most important steps to safeguard the microbiological safety of produce, washing fruits and vegetables is a common practice during processing after harvesting (Goodburn and Wallace 2013). Vigorous washing of fruits and vegetables with clean potable water typically reduces the number of microorganisms by 10- to 100-fold and is often as effective as treatment with 100 mg L$^{-1}$ chlorine, the current industry standard (Seymour and Appleton 2001).

Although chlorine is relatively cheap and easy to use, it is also highly corrosive to the stainless steel surfaces frequently used in the food industry. Its use can generate the formation of by-products, such as trihalomethanes and other chemical residues, in the wastewater, which has led to the search for new alternatives for disinfection. Therefore, most of the current investigations have been focused on the search for sanitizing treatments that can be used as alternatives to water chlorination, with the aim of assuring the quality and safety of fresh produce.

Here, the current knowledge about the efficacy of the most common decontamination procedures applied in the food industry to eliminate or inactivate HAV when present in fresh produce is summarized.

**Potable water.** Washing is an accepted step in the decontamination of fresh produce because it removes soil, pesticide residues, and some microorganisms. Croci et al. (2002) have reported less than 1-log reduction of HAV infectivity when fennel, carrots, and lettuce were washed with potable water for 5 min.

Butot et al. (2008) have evaluated the efficacy of washing for 30 s in different types of berries and vegetables, showing less than a 1.5-log reduction. Moreover, this study evaluated the effect of washing with warm water (43 °C), without any improvement on HAV elimination. Fraisse et al. (2011) also showed less than a 1-log reduction when lettuce was rinsed with water with bubbles, or bubbles and a sonication step at 35 kHz for 2 min. Overall, these results point out that rinsing with potable water had limited effects in removing HAV from berries or vegetables.

**Chlorine**, delivered as a sodium hypochlorite solution with a pH of 6.5, at concentrations of 50–200 ppm free chlorine (FC) with exposure times of 1–2 min (Goodburn and Wallace 2013), is the most widely used sanitizing agent for produce. Butot et al. (2008) measured the efficacy of washing different types of berries and vegetables with chlorinated water (200 ppm of FC). The study reported that HAV inactivation varied with the type of product, with a maximum of a 2.4-log reduction in blueberries. Chlorinated water had a limited effect on HAV titers when used to decontaminate raspberries and parsley, probably because of their surface. For instance, raspberries have crevices and hairlike projections, which may shield the viruses against environmental modifications. Nevertheless, fresh berries are unlikely to be washed because they deteriorate rapidly, unless they are used for further processing (Mariam and Cliver 2000).

Fraisse et al. (2011) have investigated the efficacy of 15 ppm of FC, determined to be the most representative of current practices after consulting a number of food manufacturers, on lettuce. They reported a 1.9-log reduction of HAV infectivity as measured by cell culture. Casteel et al. (2008) reported at least 1.2-log reductions of HAV on strawberries, tomatoes, and lettuce treated with 20 ppm chlorine for 5 min. All these results suggest that washing with chlorinated water apparently does not guarantee the complete elimination/inactivation of HAV in produce.

**Chlorine dioxide** has emerged as an alternative to chlorine since its efficacy is little affected by pH and organic matter, and it does not react with ammonia to form chloramines. However, chlorine dioxide is unstable, it must be generated on site, and it can be explosive when concentrated. Butot et al. (2008) conducted a study to explore the potential use of chlorine dioxide to decontaminate raspberries and parsley, reporting less than a 2-log reduction at 10 ppm. This concentration is more than three times the recommended concentration of 3 ppm (http://www.fda.gov/Food/default.htm).

**Peracetic acid (PAA)** is considered a potent biocide and has safe by-products of degradation (acetic acid and oxygen). Little information is available on the efficacy of PAA on HAV. To the best of my knowledge, only Fraisse et al. (2011) have reported that washing for 2 min in the presence of a peroxyacetic-based biocide (100 ppm) reduced HAV infectivity by only 0.66 log units.

**Ozone ($O_3$)** is widely used as an antimicrobial agent to disinfect water and is active against a broad range of pathogenic organisms including bacteria, protozoa, and viruses. Ozone is also one of the most effective sanitizers, with the advantage of leaving no hazardous residues on food or food-contact surfaces, and it can be used effectively in its gaseous or aqueous state (Kim et al. 2003). The inactivation of HAV by ozone has been investigated in HAV suspensions. While some tolerance to lower (i.e., 0.1–0.5 mg/L) ozone residuals was noted, the exposure of virus particles to ozone concentrations of 1 mg/L or greater at all pH levels resulted in their complete (5 log) inactivation within 60 s (Vaughn et al. 1990). Herbold et al. (1989) reported that HAV inactivation by ozone was faster at 10 °C than at 20 °C. At 20 °C, 0.25–0.38 mg of $O_3$ per liter was required for complete inactivation of HAV.

Although ozonated water has been used successfully for produce washes, reducing bacterial populations (Perry and Yousef 2011), the efficacy of ozonated water for HAV inactivation still needs to be investigated.

**Natural compounds** able to be used as biocides for washing treatments have been evaluated mainly on norovirus surrogates; however, there is not much information about their efficacy on HAV. So far, essential oils of oregano, zataria, and clove (Elizaquível et al. 2013); chitosan (Su et al. 2009); cranberry juice and cranberry proanthocyanidins (Su et al. 2010a); pomegranate juice and pomegranate polyphenols (Su et al. 2010b, 2011); black raspberry juice (Oh et al. 2012); and Korean red ginseng extract and ginsenosides (Lee et al. 2011) showed, to some extent, antiviral activity on norovirus surrogates. Until now, only grape seed extract (GSE) has been evaluated on HAV (Su and D'Souza 2011). At 0.25, 0.50, and 1 mg/mL, GSE reduced HAV infectivity by 1.81, 2.66, and 3.20 log plaque-forming units/ml, respectively, after treatment at 37 °C, in a dose-dependent manner, highlighting the potential of GSE as an inexpensive, novel, natural alternative to reduce viral contamination and enhance food safety. Lettuce and jalapeno peppers recently were inoculated with HAV and treated with several concentrations of GSE. The study reported that after 1 min, 0.25–1 mg/mL GSE caused a reduction of 0.7–1.1 and 1–1.3 log plaque-forming units for high and low HAV titers, respectively, on both food products (Su and D'Souza 2013). These results highlight the potential of natural compounds to be used as part of approaches using combinations of natural compounds with other treatments for reducing HAV on produce.

# Chapter 4
# Summary and Future Directions

The hepatitis A virus (HAV) is the etiological agent in the most common type of hepatitis worldwide, which is mainly transmitted via the fecal–oral route, either by person-to-person contact or through contaminated water and food, particularly shellfish, berries, vegetables, or ready-to-eat meals. The virus is shed in high numbers in the feces of both symptomatic and asymptomatic individuals. Even in symptomatic cases, virus shedding starts before the onset of symptoms. Since the incubation period for HAV averages 28 days, it is difficult to trace the origin of many cases and outbreaks.

Foods may be contaminated by HAV along the food supply chain, and transmission can occur by consuming food contaminated during the production process (either primary production or during further processing) by infected food handlers (Hazards 2011).

Molecular diagnostic techniques, especially polymerase chain reaction (PCR)–based methods are undoubtedly the best choice for rapid and reliable detection of HAV. In Europe, the CEN/TC25/WG6/TAG4 working group was entrusted by the European Committee for Standardization to establish a horizontal method for detecting norovirus and HAV in foods and bottled water; this methodology will be released in the coming months as an International Organization for Standardization norm (ISO/ISO/PRF TS 15216). This standard includes a real-time, reverse transcription PCR method for the detection and quantification of HAV. However, results from food testing using this standard may be disputed because it does not provide information about the infectivity of the target organism. Therefore, assessment of infectivity is tremendously important when working with food samples because of all the implications for consumers, food authorities, and the food industry.

Although the use of intercalating dyes has been investigated for discrimination of infectious HAV, much less information is available when compared with other relevant foodborne pathogens. So far, propidium monoazide combining with reverse transcription qualitative PCR has been applied successfully to discriminate between

G. Sánchez, *Hepatitis A Virus in Food: Detection and Inactivation Methods*,
SpringerBriefs in Food, Health, and Nutrition, DOI 10.1007/978-1-4614-7104-2_4,
© Gloria Sánchez 2013

infectious and noninfectious HAV suspensions (Sánchez et al. 2012). While the use of propidium monoazide already has been applied in different types of food products for the detection of other relevant foodborne pathogens (Chen et al. 2011; Elizaquível et al. 2012; Josefsen et al. 2010; Liang et al. 2011), further studies need to be undertaken to prove this concept as a method to detect infectious HAV in food samples.

The future ISO norm will bring the platform for new improvements in and developments of HAV detection methods, allowing inter- and intralaboratory standardization and validation at national and international levels. As stated above, standard methods for HAV have been adopted only recently and are not yet included in surveillance programs in many countries. Except for shellfish and fresh-cut vegetables, foods are seldom tested for viruses. Foodborne outbreaks frequently are suspected of being caused by viruses but, because of the lack of sensitive and reliable methods in most laboratories, this suspicion can rarely be confirmed by isolation of the virus from the implicated food. Hence the safety of food products cannot be assured by testing for viruses, but it can be assured by preventing contamination and implementing manufacturing processes that inactivate or eliminate them.

Koopmans and Duizer (2004) summarized the risks of consuming products that may have become contaminated with enteric viruses before processing. The risks were categorized as negligible, low, medium, and high, depending on whether the subsequent process reduced infectivity for common foodborne viruses at least 4, 3, 2, or 1 log units, respectively. To estimate risk, many studies have been devoted to evaluating the effects on HAV infectivity of treatments commonly applied to food products. Currently, there is not a single traditional or nonthermal manufacturing process alone that can claim the complete elimination or inactivation of HAV from foods. Inactivation of HAV by various food manufacturing processes differs based on the technology, mechanism of inactivation, type of food, and HAV strain. Shimasaki et al. (2009) recently demonstrated that heat and high hydrostatic pressure treatments revealed differences in inactivation efficiencies among cell-adapted HAV strains, and each strain reacted differently depending on the treatment. These differences warrant further examination, and if confirmed for other treatments, it will definitely change approaches of future inactivation studies.

Although the effect of food manufacturing technologies has been well studied in cell culture media, buffers, and some food matrices (e.g., shellfish and vegetables), there is a definite need for further research using a range of processing and technological parameters for different food products to clearly determine the conditions for efficient removal or inactivation of HAV. According to the reported data on HAV inactivation in shellfish, it is clearly shown that some of the currently used cooking processes do not inactivate HAV in lightly cooked, steamed, or sautéed molluscan shellfish. Therefore, consumers should be made aware that heating to an internal temperature of 90 °C (194 °F) for 1.5 min is required for inactivation (4-log reduction) of HAV in molluscan shellfish, even though this heat treatment may result in a less palatable product (FOODS 2008).

Often, most common food manufacturing processes by themselves are inadequate to remove or inactivate HAV in food, but when the processes are combined, the additive effect of the processes may enhance the level of HAV inactivation. Striking a balance between inactivation and the maintenance of sensory properties of these high-risk foods is an ongoing challenge to food processors.

As a consequence, emphasis should be on prevention of contamination before or during processing by implementing good agricultural, hygienic, and manufacturing practices and systems using hazard analysis critical control points. Strategies to reduce the risk of foodborne outbreaks of hepatitis A should focus on preventing foods from becoming contaminated. In developing countries, young children should be kept away from areas where fresh produce is grown and harvested. Education of workers, with an emphasis on hygiene; providing facilities for maintaining cleanliness; and the use of treated water in production and processing will be major deterrents to contamination of food with HAV. Moreover, shellfish harvesting areas should be monitored for HAV contamination.

At the consumer's level, because HAV is easily transferred between food and utensils (Wang et al. 2013), efforts have to be taken to prevent cross-contamination in the kitchen environment.

Overall, data provided in this review should serve as a baseline for food processors to effectively identify mitigation and intervention strategies in the event of an outbreak and to develop quality control measures specifically for HAV.

# References

Abad FX, Pinto RM, Bosch A (1994) Survival of enteric viruses on environmental fomites. Appl Environ Microbiol 60(10):3704–3710

Abad FX, Pinto RM, Gajardo R, Bosch A (1997) Viruses in mussels: public health implications and depuration. J Food Prot 60:677–681

Abd El Galil KH, El Sokkary MA, Kheira SM, Salazar AM, Yates MV, Chen W, Mulchandani A (2005) Real-time nucleic acid sequence-based amplification assay for detection of hepatitis A virus. Appl Environ Microbiol 71(11):7113–7116

Advisory Committee on the Microbiological Safety of Food (1998) Report on foodborne viral infections. HMSO, London

Aragonès L, Bosch A, Pintó RM (2008) Hepatitis A virus mutant spectra under the selective pressure of monoclonal antibodies: codon usage constraints limit capsid variability. J Virol 82(4):1688–1700. doi:10.1128/jvi.01842-07

Atmar RL, Neill FH, Romalde JL, Le Guyader F, Woodley CM, Metcalf TG, Estes MK (1995) Detection of Norwalk virus and hepatitis A virus in shellfish tissues with the PCR. Appl Environ Microbiol 61(8):3014–3018

Baert L, Debevere J, Uyttendaele M (2009) The efficacy of preservation methods to inactivate foodborne viruses. Int J Food Microbiol 131(2–3):83–94

Benabbes L, Ollivier J, Schaeffer J, Parnaudeau S, Rhaissi H, Nourlil J, Guyader F (2012) Norovirus and other human enteric viruses in Moroccan shellfish. Food Environ Virol 5:1–6. doi:10.1007/s12560-012-9095-8

Beuret C (2003) A simple method for isolation of enteric viruses (noroviruses and enteroviruses) in water. J Virol Methods 107:1–8

Beuret C, Kohler D, Luthi T (2000) Norwalk-like virus sequences detected by reverse transcription-polymerase chain reaction in mineral waters imported into or bottled in Switzerland. J Food Prot 63(11):1576–1582

Beuret C, Kohler D, Baumgartner A, Luthi TM (2002) Norwalk-like virus sequences in mineral waters: one-year monitoring of three brands. Appl Environ Microbiol 68(4):1925–1931

Bhattacharya SS, Kulka M, Lampel KA, Cebula TA, Goswami BB (2004) Use of reverse transcription and PCR to discriminate between infectious and non-infectious hepatitis A virus. J Virol Methods 116(2):181–187

Bianchi S, Vecchio AD, Vilariño ML, Romalde JL (2011) Evaluation of different RNA-extraction kits for sensitive detection of hepatitis A virus in strawberry samples. Food Microbiol 28(1):38–42. doi:10.1016/j.fm.2010.08.002

Bidawid S, Farber JM, Sattar SA (2000a) Contamination of foods by food handlers: experiments on hepatitis A virus transfer to food and its interruption. Appl Environ Microbiol 66(7):2759–2763

Bidawid S, Farber JM, Sattar SA (2000b) Inactivation of hepatitis A virus (HAV) in fruits and vegetables by gamma irradiation. Int J Food Microbiol 57(1–2):91–97. doi:10.1016/s0168-1605(00)00235-x

Bidawid S, Farber JM, Sattar SA (2000c) Rapid concentration and detection of hepatitis A virus from lettuce and strawberries. J Virol Methods 88(2):175–185

Bidawid S, Farber JM, Sattar SA, Hayward S (2000d) Heat inactivation of hepatitis A virus in dairy foods. J Food Prot 63(4):522–528

Bidawid S, Farber JM, Sattar SA (2001) Survival of hepatitis A virus on modified atmosphere-packaged (MAP) lettuce. Food Microbiol 18(1):95–102. doi:10.1006/fmic.2000.0380

Biziagos E, Passagot J, Crance JM, Deloince R (1988) Long-term survival of hepatitis A virus and poliovirus type 1 in mineral water. Appl Environ Microbiol 54(11):2705–2710

Blaise-Boisseau S, Hennechart-Collette C, Guillier L, Perelle S (2010) Duplex real-time qRT-PCR for the detection of hepatitis A virus in water and raspberries using the MS2 bacteriophage as a process control. J Virol Methods 166(1–2):48–53. doi:10.1016/j.jviromet.2010.02.017

Boom R, Sol CJ, Salimans MM, Jansen CL, Wertheim-van Dillen PM, van der Noordaa J (1990) Rapid and simple method for purification of nucleic acids. J Clin Microbiol 28(3):495–503

Bosch A, Xavier AF, Gajardo R, Pinto RM (1994) Should shellfish be purified before public consumption? Lancet 344(8928):1024–1025

Bosch A, Sánchez G, Le Guyader F, Vanaclocha H, Haugarreau L, Pinto RM (2001) Human enteric viruses in Coquina clams associated with a large hepatitis A outbreak. Water Sci Technol 43(12):61–65

Bosch A, Sánchez G, Abbaszadegan M, Carducci A, Guix S, Le Guyader F, Netshikweta R, Pinto RM, van der Poel W, Rutjes SA, Sano D, Taylor M, van Zyl W, Rodríguez-Lázaro D, Kovac K, Sellwood J (2011) Analytical methods for virus detection in water and food. Food Anal Methods 4(1):4–12

Bostock AD, Mepham P, Phillips S, Skidmore S, Hambling MH (1979) Hepatitis A infection associated with the consumption of mussels. J Infect 1(2):171–177. doi:10.1016/s0163-4453(79)80010-9

Boxman ILA, te Loeke NAJM, Klunder K, Hägele G, Jansen CCC (2011) Surveillance study of hepatitis A virus RNA on fig and date samples. Appl Environ Microbiol. doi:10.1128/aem.06574-11

Brassard J, Guevremont E, Gagne MJ, Lamoureux L (2011) Simultaneous recovery of bacteria and viruses from contaminated water and spinach by a filtration method. Int J Food Microbiol 144(3):565–568

Brooks HA, Gersberg RM, Dhar AK (2005) Detection and quantification of hepatitis A virus in seawater via real-time RT-PCR. J Virol Methods 127(2):109–118

Butot S, Putallaz T, Croquet C, Lamothe G, Meyer R, Joosten H, Sánchez G (2007a) Attachment of enteric viruses to bottles. Appl Environ Microbiol 73(16):5104–5110

Butot S, Putallaz T, Sánchez G (2007b) Procedure for rapid concentration and detection of enteric viruses from berries and vegetables. Appl Environ Microbiol 73(1):186–192

Butot S, Putallaz T, Sánchez G (2008) Effects of sanitation, freezing and frozen storage on enteric viruses in berries and herbs. Int J Food Microbiol 126:30–35

Butot S, Putallaz T, Amoroso R, Sánchez G (2009) Inactivation of enteric viruses in minimally processed berries and herbs. Appl Environ Microbiol 75:4155–4161

Butot S, Putallaz T, Sánchez G (2013) Improvement of procedure for HAV detection in bottled water. Food Anal Methods 6. doi:10.1007/s12161-012-9437-z

Calci KR, Meade GK, Tezloff RC, Kingsley DH (2005) High-pressure inactivation of hepatitis A virus within oysters. Appl Environ Microbiol 71(1):339–343

Calder L, Simmons G, Thornley C, Taylor P, Pritchard K, Greening G, Bishop J (2003) An outbreak of hepatitis A associated with consumption of raw blueberries. Epidemiol Infect 131(1):745–751

Cannon JL, Papafragkou E, Park GW, Osborne J, Jaykus LA, Vinje J (2006) Surrogates for the study of norovirus stability and inactivation in the environment: a comparison of murine norovirus and feline calicivirus. J Food Prot 69(11):2761–2765

Cappellozza E, Arcangeli G, Rosteghin M, Kapllan S, Magnabosco C, Bertoli E, Terregino C (2012) Survival of hepatitis A virus in pasteurized manila clams. Ital J Food Sci 24(3):247–253

Casas N, Amarita F, de Marañón IM (2007) Evaluation of an extracting method for the detection of hepatitis A virus in shellfish by SYBR-Green real-time RT-PCR. Int J Food Microbiol 120(1–2):179–185, doi:http://dx.doi.org/10.1016/j.ijfoodmicro.2007.01.017

Casteel MJ, Schmidt CE, Sobsey MD (2008) Chlorine disinfection of produce to inactivate hepatitis A virus and coliphage MS2. Int J Food Microbiol 125(3):267–273

CDC (2003) Hepatitis A outbreak associated with green onions at a restaurant–Monaca, Pennsylvania, 2003. MMWR Morb Mortal Wkly Rep 52(47):1155–1157

Chancellor DD, Tyagi S, Bazaco MC, Bacvinskas S, Chancellor MB, Dato VM, de Miguel F (2006) Green onions: potential mechanism for hepatitis A contamination. J Food Prot 69(6):1468–1472

Chen S, Wang F, Beaulieu JC, Stein RE, Ge B (2011) Rapid detection of viable salmonellae in produce by coupling propidium monoazide with loop-mediated isothermal amplification. Appl Environ Microbiol 77(12):4008–4016

Corrêa AA, Rigotto C, Moresco V, Kleemann CR, Teixeira AL, Poli CR, Simões CMO, Barardi CRM (2012) The depuration dynamics of oysters (*Crassostrea gigas*) artificially contaminated with hepatitis A virus and human adenovirus. Mem Inst Oswaldo Cruz 107(1):11–17

Costafreda MI, Bosch A, Pinto RM (2006) Development, evaluation and standardization of a real-time TaqMan reverse transcription-PCR assay for the quantification of hepatitis A virus in clinical and shellfish samples. Appl Environ Microbiol 72:3846–3855

Costa-Mattioli M, Monpoeho S, Schvoerer C, Besse B, Aleman MH, Billaudel S, Cristina J, Ferre V (2001) Genetic analysis of hepatitis A virus outbreak in France confirms the co-circulation of subgenotypes Ia, Ib and reveals a new genetic lineage. J Med Virol 65(2):233–240

Costa-Mattioli M, Cristina J, Romero H, Perez-Bercof R, Casane D, Colina R, Garcia L, Vega I, Glikman G, Romanowsky V, Castello A, Nicand E, Gassin M, Billaudel S, Ferré V (2002a) Molecular evolution of hepatitis A virus: a new classification based on the complete VP1 protein. J Virol 76:9516–9525

Costa-Mattioli M, Monpoeho S, Nicand E, Aleman MH, Billaudel S, Ferre V (2002b) Quantification and duration of viraemia during hepatitis A infection as determined by real-time RT-PCR. J Viral Hepat 9(2):101–106

Cotter SM, Sansom S, Long T, Koch E, Kellerman S, Smith F, Averhoff F, Bell BP (2003) Outbreak of Hepatitis A among Men Who Have Sex with Men: Implications for Hepatitis A Vaccination Strategies. Journal of Infectious Diseases 187:1235–1240 doi:10.1086/374057

Croci L, Ciccozzi M, De Medici D, Di Pasquale S, Fiore A, Mele A, Toti L (1999) Inactivation of hepatitis A virus in heat-treated mussels. J Appl Microbiol 87(6):884–888. doi:10.1046/j.1365-2672.1999.00935.x

Croci L, De MD, Scalfaro C, Fiore A, Toti L (2002) The survival of hepatitis A virus in fresh produce. Int J Food Microbiol 73(1):29–34

Croci L, de Medici D, Di Pasquale S, Toti L (2005) Resistance of hepatitis A virus in mussels subjected to different domestic cookings. Int J Food Microbiol 105(2):139–144

Croci L, Dubois E, Cook N, de Medici D, Schulz AC, China B, Rutjes SA, Hoofar J, van der Poel WH (2008) Current methods for extraction and concentration of enteric viruses from fresh fruit and vegetables: towards international standards. Food Anal Methods 1:73–84

D'Agostino M, Cook N, Rodriguez-Lazaro D, Rutjes S (2011) Nucleic acid amplification-based methods for detection of enteric viruses: definition of controls and interpretation of results. Food Environ Virol 3:55–60

Dagan R, Leventhal A, Anis E, Slater P, Ashur Y, Shouval D (2005) Incidence of hepatitis a in israel following universal immunization of toddlers. JAMA 294(2):202–210. doi:10.1001/jama.294.2.202

De Abreu Corrêa A, Souza DSM, Moresco V, Kleemann CR, Garcia LAT, Barardi CRM (2012) Stability of human enteric viruses in seawater samples from mollusc depuration tanks coupled with ultraviolet irradiation. J Appl Microbiol 113(6):1554–1563. doi:10.1111/jam.12010

De Medici D, Ciccozzi M, Fiore A, Di Pasquale S, Parlato A, Ricci-Bitti P, Croci L (2001) Closed-circuit system for the depuration of mussels experimentally contaminated with hepatitis A virus. J Food Prot 64(6):877–880

Deboosere N, Legeay O, Caudrelier Y, Lange M (2004) Modelling effect of physical and chemical parameters on heat inactivation kinetics of hepatitis A virus in a fruit model system. Int J Food Microbiol 93(1):73–85

Deboosere N, Pinon A, Delobel A, Temmam S, Morin T, Merle G, Blaise-Boisseau S, Perelle S, Vialette M (2010) A predictive microbiology approach for thermal inactivation of hepatitis A virus in acidified berries. Food Microbiol 27(7):962–967, doi:http://dx.doi.org/10.1016/j.fm.2010.05.018

Dentinger CM, Bower WA, Nainan OV, Cotter SM, Myers G, Dubusky LM, Fowler S, Salehi ED, Bell BP (2001) An outbreak of hepatitis A associated with green onions. J Infect Dis 183(8):1273–1276

Di Pasquale S, Paniconi M, Auricchio B, Orefice L, Schultz AC, de Medici D (2010a) Comparison of different concentration methods for the detection of hepatitis A virus and calicivirus from bottled natural mineral waters. J Virol Methods 165(1):57–63

Di Pasquale S, Paniconi M, De Medici D, Suffredini E, Croci L (2010b) Duplex real time PCR for the detection of hepatitis A virus in shellfish using feline calicivirus as a process control. J Virol Methods 163(1):96–100. doi:10.1016/j.jviromet.2009.09.003

Domínguez A, Oviedo M, Carmona G, Batalla J, Bruguera M, Salleras L, Plasència A (2008) Impact and effectiveness of a mass hepatitis A vaccination programme of preadolescents seven years after introduction. Vaccine 26(14):1737–1741, doi:http://dx.doi.org/10.1016/j.vaccine.2008.01.048

Doyle MP, Erickson MC (2008) Summer meeting 2007 – the problems with fresh produce: an overview. J Appl Microbiol 105(2):317–330

Dubois E, Agier C, Traore O, Hennechart C, Merle G, Cruciere C, Laveran H (2002) Modified concentration method for the detection of enteric viruses on fruits and vegetables by reverse transcriptase-polymerase chain reaction or cell culture. J Food Prot 65(12):1962–1969

Dubois E, Hennechart C, Deboosere N, Merle G, Legeay O, Burger C, Le Calve M, Lombard B, Ferre V, Traore O (2006) Intra-laboratory validation of a concentration method adapted for the enumeration of infectious F-specific RNA coliphage, enterovirus, and hepatitis A virus from inoculated leaves of salad vegetables. Int J Food Microbiol 108(2):164–171

Ehlers MM, Van Zyl WB, Pavlov DN, Muller EE (2004) Random survey of the microbial quality of bottled water in South Africa. Water SA 30(2):203–210

Elizaquível P, Sánchez G, Aznar R (2012) Quantitative detection of viable foodborne E. coli O157:H7, Listeria monocytogenes and Salmonella in fresh-cut vegetables combining propidium monoazide and real-time PCR. Food Control 25(2):704–708. doi:10.1016/j.foodcont.2011.12.003

Elizaquível PA, Azizkhani M, Aznar R, Sánchez G (2013) The effect of essential oils on norovirus surrogates. Food Control 32:275–278

Fauquet CM, Mayo MA, Maniloff J, Desselberger U, Ball LA (2005) Virus taxonomy: the eighth report of the International Committee on Taxonomy of Viruses. Elsevier Academic Press, Amsterdam

Felix-Valenzuela L, Resendiz-Sandoval M, Burgara-Estrella A, Hernández J, Mata-Haro V (2012) Quantitative detection of hepatitis A, rotavirus and genogroup I norovirus by RT-qPCR in fresh produce from packinghouse facilities. J Food Saf 32(4):467–473

Fino VR, Kniel KE (2008a) Comparative recovery of foodborne viruses from fresh produce. Foodborne Pathog Dis 5(6):819–825

Fino VR, Kniel KE (2008b) UV light inactivation of hepatitis A virus, Aichi virus, and feline calicivirus on strawberries, green onions, and lettuce. J Food Prot 71(5):908–913

Fiore AE (2004) Hepatitis A transmitted by food. Clin Infect Dis 38:705–715

FOODS NACOMCF (2008) Response to the questions posed by the Food and Drug Administration and the National Marine Fisheries Service regarding determination of cooking parameters for safe seafood for consumers. J Food Prot 71:1287–1308

Fraisse A, Temmam S, Deboosere N, Guillier L, Delobel A, Maris P, Vialette M, Morin T, Perelle S (2011) Comparison of chlorine and peroxyacetic-based disinfectant to inactivate feline calicivirus, murine norovirus and hepatitis A virus on lettuce. Int J Food Microbiol 151(1): 98–104. doi:10.1016/j.ijfoodmicro.2011.08.011

Franco E, Vitiello G (2003) Vaccination strategies against hepatitis A in southern Europe. Vaccine 21(7–8):696–697, doi:http://dx.doi.org/10.1016/S0264-410X(02)00582-0

Frank C, Walter J, Muehlen M, Jansen A, van Treeck U, Hauri AM, Zoellner I, Rakha M, Hoehne M, Hamouda O, Schreier E, Stark K (2007) Major outbreak of hepatitis A associated with orange juice among tourists, Egypt, 2004. Emerg Infect Dis 13(1):156–158. doi:10.3201/eid1301.060487

Furuta T, Akiyama M, Kato Y, Nishio O (2003) A food poisoning outbreak caused by purple Washington clam contaminated with norovirus (Norwalk-like virus) and hepatitis A virus. Kansenshogaku Zasshi 77:89–94

Gallot C, Grout L, Roque-Afonso AM, Couturier E, Carrillo-Santisteve P, Pouey J, Letort MJ, Hoppe S, Capdepon P, Saint-Martin S, De Valk H, Vaillant V (2011) Hepatitis A associated with semidried tomatoes, France, 2010. Emerg Infect Dis 17(3):566–567. doi:10.3201/eid1703.101479

Gilgen M, Germann D, Lüthy J, Hübner P (1997) Three-step isolation method for sensitive detection of enterovirus, rotavirus, hepatitis A virus, and small round structured viruses in water samples. Int J Food Microbiol 37(2–3):189–199

Goh KT, Chan L, Ding JL, Oon CJ (1984) An epidemic of cockles-associated hepatitis A in Singapore. Bull World Health Organ 62(6):893–897

Goodburn C, Wallace CA (2013) The microbiological efficacy of decontamination methodologies for fresh produce: a review. Food Control. 32:418–427 doi:10.1016/j.foodcont.2012.12.012

Guevremont E, Brassard J, Houde A, Simard C, Trottier YL (2006) Development of an extraction and concentration procedure and comparison of RT-PCR primer systems for the detection of hepatitis A virus and norovirus GII in green onions. J Virol Methods 134(1–2):130–135

Halliday ML, Kang L-Y, Zhou T-K, Hu M-D, Pan Q-C, Fu T-Y, Huang Y-S, Hu S-L (1991) An epidemic of hepatitis A attributable to the ingestion of raw clams in Shanghai, China. J Infect Dis 164(5):852–859. doi:10.2307/30111993

Haughton PN, Lyng JG, Cronin DA, Morgan DJ, Fanning S, Whyte P (2012) Efficacy of pulsed electric fields for the inactivation of indicator microorganisms and foodborne pathogens in liquids and raw chicken. Food Control 25(1):131–135

Hazards EPoB (2011) Scientific Opinion on an update on the present knowledge on the occurrence and control of foodborne viruses. EFSA J 9:2190. doi:10.2903/j.efsa.2011.2190

Heitmann A, Laue T, Schottstedt V, Dotzauer A, Pichl L (2005) Occurrence of hepatitis A virus genotype III in Germany requires the adaptation of commercially available diagnostic test systems. Transfusion 45(7):1097–1105

Herbold K, Flehmig B, Botzenhart K (1989) Comparison of ozone inactivation, in flowing water, of hepatitis A virus, poliovirus 1, and indicator organisms. Appl Environ Microbiol 55(11):2949–2953

Hernandez F, Monge R, Jimenez C, Taylor L (1997) Rotavirus and hepatitis A virus in market lettuce (Latuca sativa) in Costa Rica. Int J Food Microbiol 37:221–223

Hewitt J, Greening GE (2004) Survival and persistence of norovirus, hepatitis A virus, and feline calicivirus in marinated mussels. J Food Prot 67(8):1743–1750

Hewitt J, Greening GE (2006) Effect of heat treatment on hepatitis A virus and norovirus in New Zealand greenshell mussels (Perna canaliculus) by quantitative real-time reverse transcription PCR and cell culture. J Food Prot 69(9):2217–2223

Hida K, Kulka M, Papafragkou E (2013) Development of a rapid total nucleic acid extraction method for the isolation of hepatitis A virus from fresh produce. Int J Food Microbiol 161:143–150, doi:http://dx.doi.org/10.1016/j.ijfoodmicro.2012.12.007

Hirneisen K, Kniel K (2012) Comparative uptake of enteric viruses into spinach and green onions. Food Environ Virol 5:1–11 doi:10.1007/s12560-012-9093-x

Hollinger FB, Emerson SU (2007) Hepatitis A virus. In: Knipe DN, Howley PM (eds) Fields virology, 5th edn. Lippincott Williams & Wilkins, Philadelphia, pp 911–947

Houde A, Guévremont E, Poitras E, Leblanc D, Ward P, Simard C, Trottier Y-L (2007) Comparative evaluation of new TaqMan real-time assays for the detection of hepatitis A virus. J Virol Methods 140(1–2):80–89. doi:10.1016/j.jviromet.2006.11.003

Hu Y, Arsov I (2009) Nested real-time PCR for hepatitis A detection. Lett Appl Microbiol 49(5):615–619. doi:10.1111/j.1472-765X.2009.02713.x

Hutin YJ, Pool V, Cramer EH, Nainan OV, Weth J, Williams IT, Goldstein ST, Gensheimer KF, Bell BP, Shapiro CN, Alter MJ, Margolis HS (1999) A multistate, foodborne outbreak of hepatitis A. N Engl J Med 340(8):595–602

Hyeon J-Y, Chon J-W, Park C, Lee J-B, Choi I-S, Kim M-S, Seo K-H (2011) Rapid detection method for hepatitis A virus from lettuce by a combination of filtration and integrated cell culture-real-time reverse transcription PCR. J Food Prot 74(10):1756–1761. doi:10.4315/0362-028x.jfp-11-155

Jaykus LA, De Leon R, Sobsey MD (1995) Development of a molecular method for the detection of enteric viruses in oysters. J Food Prot 58(12):1357–1362

Jean J, Blais B, Darveau A, Fliss I (2001) Detection of hepatitis A virus by the nucleic acid sequence-based amplification technique and comparison with reverse transcription-PCR. Appl Environ Microbiol 67(12):5593–5600

Josefsen MH, Lofstrom C, Hansen TB, Christensen LS, Olsen JE, Hoorfar J (2010) Rapid quantification of viable Campylobacter bacteria on chicken carcasses, using real-time PCR and propidium monoazide treatment, as a tool for quantitative risk assessment. Appl Environ Microbiol 76(15):5097–5104

Jothikumar N, Cromeans TL, Sobsey MD, Robertson BH (2005) Development and evaluation of a broadly reactive TaqMan assay for rapid detection of hepatitis A virus. Appl Environ Microbiol 71(6):3359–3363

Khanna N, Goldenberger D, Graber P, Battegay M, Widmer AF (2003) Gastroenteritis outbreak with norovirus in a Swiss university hospital with a newly identified virus strain. J Hosp Infect 55(2):131–136

Kim JG, Yousef AE, Khadre MA (2003) Ozone and its current and future application in the food industry. Adv Food Nutr Res 45:167–218

Kingsley DH, Richards GP (2001) Rapid and efficient extraction method for reverse transcription-PCR detection of hepatitis A and Norwalk-like viruses in shellfish. Appl Environ Microbiol 67(9):4152–4157

Kingsley DH, Hoover DG, Papafragkou E, Richards GP (2002) Inactivation of hepatitis A virus and a calicivirus by high hydrostatic pressure. J Food Prot 65(10):1605–1609

Kingsley DH, Guan D, Hoover DG (2005) Pressure inactivation of hepatitis A virus in strawberry puree and sliced green onions. J Food Prot 68(8):1748–1751

Koff RS, Sear HS (1967) Internal temperature of steamed clams. N Engl J Med 276(13):737–739. doi:10.1056/NEJM196703302761307

Kokkinos P, Kozyra I, Lazic S, Bouwknegt M, Rutjes S, Willems K, Moloney R, Roda Husman AM, Kaupke A, Legaki E, D'Agostino M, Cook N, Rzeżutka A, Petrovic T, Vantarakis A (2012) Harmonised investigation of the occurrence of human enteric viruses in the leafy green vegetable supply chain in three European countries. Food Environ Virol 4:1–13 doi:10.1007/s12560-012-9087-8

Konduru K, Kaplan GG (2006) Stable growth of wild-type hepatitis A virus in cell culture. J Virol 80(3):1352–1360

Konowalchuk J, Speirs JI (1975) Survival of enteric viruses on fresh fruit. J Milk Food Technol 38(598):600

Koopmans M, Duizer E (2004) Foodborne viruses: an emerging problem. Int J Food Microbiol 90(1):23–41

Kovac K, Gutierrez-Aguirre I, Banjac M, Peterka M, Poljsak-Prijatelj M, Ravnikar M, Mijovski JZ, Schultz AC, Raspor P (2009) A novel method for concentrating hepatitis A virus and caliciviruses from bottled water. J Virol Methods 162(1–2):272–275

Kovač K, Diez-Valcarce M, Hernandez M, Raspor P, Rodríguez-Lázaro D (2010) High hydrostatic pressure as emergent technology for the elimination of foodborne viruses. Trends Food Sci Technol 21(11):558–568. doi:10.1016/j.tifs.2010.08.002

Laird DT, Sun Y, Reineke KF, Carol Shieh Y (2011) Effective hepatitis A virus inactivation during low-heat dehydration of contaminated green onions. Food Microbiol 28(5):998–1002. doi:10.1016/j.fm.2011.01.011

Lamothe GT, Putallaz T, Joosten H, Marugg JD (2003) Reverse transcription-PCR analysis of bottled and natural mineral waters for the presence of noroviruses. Appl Environ Microbiol 69(11):6541–6549

Lee MHL, Lee BH, Jung JY, Cheon DS, Kim KT, Choi C (2011) Antiviral effect of Korean red ginseng extract and ginsenosides on murine norovirus and feline calicivirus as surrogates for human norovirus. J Ginseng Res 35(4):429–435

Lee K, Lee H, Ha S-D, Cheon D-S, Choi C (2012) Comparative analysis of viral concentration methods for detecting the HAV genome using real-time RT-PCR amplification. Food Environ Virol 4(2):68–72. doi:10.1007/s12560-012-9077-x

Lees D (2000) Viruses and bivalve shellfish. Int J Food Microbiol 59(1–2):81–116, doi:http://dx.doi.org/10.1016/S0168-1605(00)00248-8

Lees D (2010) International standardisation of a method for detection of human pathogenic viruses in molluscan shellfish. Food Environ Virol 2(3):146–155. doi:10.1007/s12560-010-9042-5

Leggitt PR, Jaykus LA (2000) Detection methods for human enteric viruses in representative foods. J Food Prot 63(12):1738–1744

Lemon SM, Binn LN (1983) Antigenic relatedness of two strains of hepatitis A virus determined by cross-neutralization. Infect Immun 42(1):418–420

Li JW, Xin ZT, Wang XW, Zheng JL, Chao FH (2002) Mechanisms of inactivation of hepatitis A virus by chlorine. Appl Environ Microbiol 68(10):4951–4955. doi:10.1128/aem.68.10.4951-4955.2002

Liang N, Dong J, Luo L, Li Y (2011) Detection of viable Salmonella in lettuce by propidium monoazide real-time PCR. J Food Sci 76(4):M234–M237

Liu J, Wu Q, Kou X (2007) Development of a virus concentration method and its application for the detection of noroviruses in drinking water in China. J Microbiol 45(1):48–52

Love DC, Casteel MJ, Meschke JS, Sobsey MD (2008) Methods for recovery of hepatitis A virus (HAV) and other viruses from processed foods and detection of HAV by nested RT-PCR and TaqMan RT-PCR. Int J Food Microbiol 126(1–2):221–226. doi:10.1016/j.ijfoodmicro.2008.05.032

Lu L, Ching KZ, De Paula VS, Nakano T, Siegl G, Weitz M, Robertson BH (2004) Characterization of the complete genomic sequence of genotype II hepatitis A virus (CF53/Berne isolate). J Gen Virol 85:2943–2952

Lynch MF, Tauxe RV, Hedberg CW (2009) The growing burden of foodborne outbreaks due to contaminated fresh produce: risks and opportunities. Epidemiol Infect 137(Special Issue 03):307–315

Mackowiak PA, Caraway CT, Portnoy BL (1976) Oyster-associated hepatitis: lessons from the louisiana experience. Am J Epidemiol 103(2):181–191

Mallett JC, Beghian LE, Metcalf TG, Kaylor JD (1991) Potential of irradiation technology for improved shellfish sanitation. J Food Saf 11(4):231–245. doi:10.1111/j.1745-4565.1991.tb00055.x

Mariam TW, Cliver DO (2000) Hepatitis A virus control in strawberry products. Dairy Food Environ Sanit 20(8):612–616

Martínez-Abad A, Ocio MJ, Lagarón JM, Sánchez G (2013) Evaluation of silver-infused polylactide films for inactivation of Salmonella and feline calicivirus in vitro and on fresh-cut vegetables. Int J Food Microbiol 162:89–94, doi:http://dx.doi.org/10.1016/j.ijfoodmicro.2012.12.024

Martin-Latil S, Hennechart-Collette C, Guillier L, Perelle S (2012) Comparison of two extraction methods for the detection of hepatitis A virus in semi-dried tomatoes and murine norovirus as a process control by duplex RT-qPCR. Food Microbiol 31(2):246–253. doi:10.1016/j. fm.2012.03.007

Mattison K, Brassard J, Gagne MJ, Ward P, Houde A, Lessard L, Simard C, Shukla A, Pagotto F, Jones TH, Trottier YL (2009) The feline calicivirus as a sample process control for the detection of food and waterborne RNA viruses. Int J Food Microbiol 132(1):73–77

Mbithi JN, Springthorpe VS, Sattar SA (1991) Effect of relative humidity and air temperature on survival of hepatitis A virus on environmental surfaces. Appl Environ Microbiol 57(5):1394–1399

Milazzo L, Vale S (2005) Hepatitis A associated with green onions. New Engl J Med 353(21):2300–2301

Millard J, Appleton H, Parry JV (1987) Studies on heat inactivation of hepatitis A virus with special reference to shellfish. Part 1. Procedures for infection and recovery of virus from laboratory-maintained cockles. Epidemiol Infect 98(3):397–414

Monge R, Arias ML (1996) Occurence of some pathogenic microorganisms in fresh vegetables in Costa Rica. Arch Latinoam Nutr 46(4):292–294

Morales-Rayas R, Wolffs PFG, Griffiths MW (2010) Simultaneous separation and detection of hepatitis A virus and norovirus in produce. Int J Food Microbiol 139(1–2):48–55. doi:10.1016/j. ijfoodmicro.2010.02.011

Mukhopadhyay S, Ramaswamy R (2012) Application of emerging technologies to control Salmonella in foods: a review. Food Res Int 45(2):666–677

Nainan OV, Xia G, Vaughan G, Margolis HS (2006) Diagnosis of hepatitis A virus infection: a molecular approach. Clin Microbiol Rev 19(1):63–79

Niu MT, Polish LB, Robertson BH, Khanna BK, Woodruff BA, Shapiro CN, Miller MA, Smith JD, Gedrose JK, Alter MJ, Margolis HS (1992) Multistate outbreak of hepatitis A associated with frozen strawberries. J Infect Dis 166(3):518–524

Noble RC, Kane MA, Reeves SA, Roeckel I (1984) Posttransfusion hepatitis A in a neonatal intensive care unit. JAMA 252(19):2711–2725

Nuanualsuwan S, Cliver DO (2003) Capsid functions of inactivated human picornaviruses and feline calicivirus. Appl Environ Microbiol 69(1):350–357

Oh M, Bae SY, Lee JH, Cho KJ, Kim KH, Chung MS (2012) Antiviral effects of black raspberry (*Rubus coreanus*) juice on foodborne viral surrogates. Foodborne Pathog Dis 9(10):915–921. doi:10.1089/fpd.2012.1174

Papafragkou E, Plante M, Mattison K, Bidawid S, Karthikeyan K, Farber JM, Jaykus L-A (2008) Rapid and sensitive detection of hepatitis A virus in representative food matrices. J Virol Methods 147(1):177–187. doi:10.1016/j.jviromet.2007.08.024

Parshionikar S, Laseke I, Fout GS (2010) Use of propidium monoazide in reverse transcriptase PCR to distinguish between infectious and noninfectious enteric viruses in water samples. Appl Environ Microbiol 76(13):4318–4326

Paula VS, Gaspar AMC, Villar LM (2010) Optimization of methods for detecting hepatitis A virus in food. Food Environ Virol 2(1):47–52. doi:10.1007/s12560-010-9027-4

Pebody RG, Leino T, Ruutu P, Kinnunen L, Davidkin I, Nohynek H, Leinikki P (1998) Foodborne outbreaks of hepatitis A in a low endemic country: an emerging problem? Epidemiol Infect 120(1):55–59

Perelle S, Cavellini L, Burger C, Blaise-Boisseau S, Hennechart-Collette C, Merle G, Fach P (2009) Use of a robotic RNA purification protocol based on the NucliSens® easyMAG(TM) for real-time RT-PCR detection of hepatitis A virus in bottled water. J Virol Methods 157(1):80–83

Pérez-Sautu U, Costafreda MI, Lite J, Sala R, Barrabeig I, Bosch A, Pintó RM (2011) Molecular epidemiology of hepatitis A virus infections in Catalonia, Spain, 2005–2009: circulation of newly emerging strains. J Clin Virol 52(2):98–102. doi:10.1016/j.jcv.2011.06.011

Perry JJ, Yousef AE (2011) Decontamination of raw foods using ozone-based sanitization techniques. Annu Rev Food Sci Technol 2:281–298

Petrignani M, Harms M, Verhoef L, van Hunen R, Swaan C, van Steenbergen J, Boxman I, Sala RPI, Ober HJ, Vennema H, Koopmans M, van Pelt W (2010) Update: a food-borne outbreak of hepatitis a in the Netherlands related to semi-dried tomatoes in oil, January-February 2010. Eurosurveillance 15(20)

Pinto RM, Alegre D, Dominguez A, El Senousy WM, Sánchez G, Villena C, Costafreda MI, Aragones L, Bosch A (2007) Hepatitis A virus in urban sewage from two Mediterranean countries. Epidemiol Infect 135:270–273

Pinto RM, Costafreda MI, Bosch A (2009) Risk assessment in shellfish-borne outbreaks of hepatitis A. Appl Environ Microbiol 75(23):7350–7355

Prato R, Martinelli D, Tafuri S, Barbuti G, Quarto M, Germinario CA, Chironna M (2013) Safety of shellfish and epidemiological pattern of enterically transmitted diseases in Italy. Int J Food Microbiol 162:125–128. doi:http://dx.doi.org/10.1016/j.ijfoodmicro.2012.12.025

Puértolas E, Álvarez I, Raso J, Martínez de Marañón I (2012) Industrial application of pulsed electric field for food pasteurization: review of its technical and commercial viability. CyTA J Food 11:1–8. doi:10.1080/19476337.2012.693542

Ramsay CN, Upton PA (1989) Hepatitis A and frozen raspberries. Lancet 1(8628):43–44

Reid TM, Robinson HG (1987) Frozen raspberries and hepatitis A. Epidemiol Infect 98(1):109–112

Ribao C, Torrado I, Vilarino ML, Romalde JL (2004) Assessment of different commercial RNA-extraction and RT-PCR kits for detection of hepatitis A virus in mussel tissues. J Virol Methods 115:177–182

Richards G, McLeod C, Guyader F (2010) Processing strategies to inactivate enteric viruses in shellfish. Food Environ Virol 2(3):183–193. doi:10.1007/s12560-010-9045-2

Robertson BH, Jansen RW, Khanna B, Tosuka A, Nainan OV, Siegl G, Widell A, Margolis HS, Isomura S, Ito K, Ishizu T, Moritsugu Y, Lemon SM (1992) Genetic relatedness of hepatitis A virus strains recovered from different geographical regions. J Gen Virol 73:1365–1377

Robesyn E, De Schrijver K, Wollants E, Top G, Verbeeck J, Van Ranst M (2009) An outbreak of hepatitis A associated with the consumption of raw beef. J Clin Virol 44(3):207–210. doi:10.1016/j.jcv.2008.12.012

Rodrigo D, van Loey A, Hendrickx M (2007) Combined thermal and high pressure colour degradation of tomato puree and strawberry juice. J Food Eng 79(2):553–560. doi:10.1016/j.jfoodeng.2006.02.015

Rodriguez RA, Pepper IL, Gerba CP (2009) Application of PCR-based methods to assess the infectivity of enteric viruses in environmental samples. Appl Environ Microbiol 75(2):297–307

Romalde JL, Estes MK, Szucs G, Atmar RL, Woodley CM, Metcalf TG (1994) In situ detection of hepatitis A virus in cell cultures and shellfish tissues. Appl Environ Microbiol 60(6):1921–1926

Rosenblum LS, Mirkin IR, Allen DT, Safford S, Hadler SC (1990) A multifocal outbreak of hepatitis A traced to commercially distributed lettuce. Am J Public Health 80(9):1075–1079

Rzezutka A, Cook N (2004) Survival of human enteric viruses in the environment and food. FEMS Microbiol Rev 28:441–453

Rzezutka A, D'Agostino M, Cook N (2006) An ultracentrifugation-based approach to the detection of hepatitis A virus in soft fruits. Int J Food Microbiol 108(3):315–320

Sair AI, D'Souza DH, Moe CL, Jaykus LA (2002) Improved detection of human enteric viruses in foods by RT-PCR. J Virol Methods 100(1–2):57–69

Sánchez G, Pinto RM, Vanaclocha H, Bosch A (2002) Molecular characterization of hepatitis a virus isolates from a transcontinental shellfish-borne outbreak. J Clin Microbiol 40:4148–4155

Sánchez G, Bosch A, Gomez-Mariano G, Domingo E, Pinto RM (2003a) Evidence for quasispecies distributions in the human hepatitis A virus genome. Virology 315(1):34–42

Sánchez G, Bosch A, Pinto RM (2003b) Genome variability and capsid structural constraints of hepatitis a virus. J Virol 77(1):452–459

Sánchez G, Joosten H, Meyer R (2005) Presence of norovirus sequences in bottled waters is questionable. Appl Environ Microbiol 71(4):2203–2205

Sánchez G, Populaire S, Butot S, Putallaz T, Joosten H (2006) Detection and differentiation of human hepatitis A strains by commercial quantitative real-time RT-PCR tests. J Virol Methods 132(1–2):160–165

Sánchez G, Bosch A, Pinto RM (2007) Hepatitis A virus detection in food: current and future prospects. Lett Appl Microbiol 45:1–5

Sánchez G, Elizaquivel P, Aznar R (2012a) Discrimination of infectious hepatitis A viruses by propidium monoazide real-time RT-PCR. Food Environ Virol 4:21–25

Sánchez G, Elizaquível P, Aznar R (2012b) A single method for recovery and concentration of enteric viruses and bacteria from fresh-cut vegetables. Int J Food Microbiol 152(1–2):9–13. doi:10.1016/j.ijfoodmicro.2011.10.002

Sattar SA, Jason T, Bidawid S, Farber J (2000) Foodborne spread of hepatitis A: recent studies on virus survival, transfer and inactivation. Can J Infect Dis 11(3):159–163

Scharff RL (2012) Economic burden from health losses due to foodborne illness in the United States. J Food Prot 75(1):123–131. doi:10.4315/0362-028x.jfp-11-058

Schmid D, Fretz R, Buchner G, König C, Perner H, Sollak R, Tratter A, Hell M, Maass M, Strasser M, Allerberger F (2009) Foodborne outbreak of hepatitis A, November 2007–January 2008, Austria. Eur J Clin Microbiol Infect Dis 28(4):385–391. doi:10.1007/s10096-008-0633-0

Scholz E, Heinricy U, Flehmig B (1989) Acid stability of hepatitis A virus. J Gen Virol 70(9):2481–2485. doi:10.1099/0022-1317-70-9-2481

Schultz AC, Perelle S, Di PS, Kovac K, De MD, Fach P, Sommer HM, Hoorfar J (2011) Collaborative validation of a rapid method for efficient virus concentration in bottled water. Int J Food Microbiol 145(Suppl 1):S158–S166

Seymour IJ, Appleton H (2001) Foodborne viruses and fresh produce. J Appl Microbiol 91(5):759–773

Shan XC, Wolffs P, Griffiths MW (2005) Rapid and quantitative detection of hepatitis A virus from green onion and strawberry rinses by use of real-time reverse transcription-PCR. Appl Environ Microbiol 71(9):5624–5626

Sharma M, Shearer AEH, Hoover DG, Liu MN, Solomon MB, Kniel KE (2008) Comparison of hydrostatic and hydrodynamic pressure to inactivate foodborne viruses. Innov Food Sci Emerg Technol 9(4):418–422

Shieh YC, Khudyakov YE, Xia G, Ganova-Raeva LM, Khambaty FM, Woods JW, Veazey JE, Motes ML, Glatzer MB, Bialek SR, Fiore AE (2007) Molecular confirmation of oysters as the vector for hepatitis A in a 2005 multistate outbreak. J Food Prot 70(1):145–150

Shieh YC, Stewart DS, Laird DT (2009) Survival of hepatitis A virus in spinach during low temperature storage. J Food Prot 72(11):2390–2393

Shimasaki N, Kiyohara T, Totsuka A, Nojima K, Okada Y, Yamaguchi K, Kajioka J, Wakita T, Yoneyama T (2009) Inactivation of hepatitis A virus by heat and high hydrostatic pressure: variation among laboratory strains. Vox Sang 96(1):14–19. doi:10.1111/j.1423-0410.2008.01113.x

Silberstein E, Xing L, van de Beek W, Lu J, Cheng H, Kaplan GG (2003) Alteration of hepatitis A virus (HAV) particles by a soluble form of HAV cellular receptor 1 containing the immuno-globulin- and mucin-like regions. J Virol 77(16):8765–8774

Sincero TCM, Levin DB, Simões CMO, Barardi CRM (2006) Detection of hepatitis A virus (HAV) in oysters (*Crassostrea gigas*). Water Res 40(5):895–902, doi:http://dx.doi.org/10.1016/j.watres.2005.12.005

Sow H, Desbiens M, Morales-Rayas R, Ngazoa SE, Jean J (2011) Heat inactivation of hepatitis A virus and a norovirus surrogate in soft-shell clams (*Mya arenaria*). Foodborne Pathog Dis 8(3):387–393

Stals A, Baert L, Van Coillie E, Uyttendaele M (2012) Extraction of food-borne viruses from food samples: a review. Int J Food Microbiol 153(1–2):1–9

Stine SW, Song I, Choi CY, Gerba CP (2005) Effect of relative humidity on preharvest survival of bacterial and viral pathogens on the surface of cantaloupe, lettuce, and bell peppers. J Food Prot 68(7):1352–1358

Su X, D'Souza DH (2011) Grape seed extract for control of human enteric viruses. Appl Environ Microbiol 77(12):3982–3987

Su X, D'Souza DH (2013) Grape seed extract for foodborne virus reduction on produce. Food Microbiol 34:1–6. doi:10.1016/j.fm.2012.10.006

Su X, Zivanovic S, D'Souza DH (2009) Effect of chitosan on the infectivity of murine norovirus, feline calicivirus, and bacteriophage MS2. J Food Prot 72(12):2623–2628

Su X, Howell AB, D'Souza DH (2010a) Antiviral effects of cranberry juice and cranberry proanthocyanidins on foodborne viral surrogates–a time dependence study in vitro. Food Microbiol 27(8):985–991

Su X, Sangster MY, D'Souza DH (2010b) In vitro effects of pomegranate juice and pomegranate polyphenols on foodborne viral surrogates. Foodborne Pathog Dis 7(12):1473–1479

Su X, Sangster MY, D'Souza DH (2011) Time-dependent effects of pomegranate juice and pomegranate polyphenols on foodborne viral reduction. Foodborne Pathog Dis 8(11):1177–1183. doi:10.1089/fpd.2011.0873

Sun Y, Laird DT, Shieh YC (2012) Temperature-dependent survival of hepatitis A virus during storage of contaminated onions. Appl Environ Microbiol 78(14):4976–4983. doi:10.1128/aem.00402-12

Terio V, Tantillo G, Martella V, Pinto P, Buonavoglia C, Kingsley D (2010) High pressure inactivation of HAV within mussels. Food Environ Virol 2:83–88

Tortajada C, de Olalla P, Diez E, Pinto R, Bosch A, Perez U, Sanz M, Cayla J, Group SW (2012) Hepatitis a among men who have sex with men in Barcelona, 1989–2010: insufficient control and need for new approaches. BMC Infectious Diseases 12:11

Van Boxstael S, Habib I, Jacxsens L, De Vocht M, Baert L, Van De Perre E, Rajkovic A, Lopez-Galvez F, Sampers I, Spanoghe P, De Meulenaer B, Uyttendaele M (2013) Food safety issues in fresh produce: bacterial pathogens, viruses and pesticide residues indicated as major concerns by stakeholders in the fresh produce chain. Food Control 32:190–197

Vaughn JM, Chen YS, Novotny JF, Strout D (1990) Effects of ozone treatment on the infectivity of hepatitis A virus. Can J Microbiol 36:557–560

Villar LM, De Paula VS, Diniz-Mendes L, Lampe E, Gaspar AM (2006) Evaluation of methods used to concentrate and detect hepatitis A virus in water samples. J Virol Methods 137(2):169–176

Wang MJ, Moran GJ (2004) Hepatitis A outbreak associated with green onions at a restaurant – Monaca, Pennsylvania, 2003. Ann Emerg Med 43(5):660–663

Wang Q, Erickson M, Ortega YR, Cannon JL (2013) The fate of murine norovirus and hepatitis A virus during preparation of fresh produce by cutting and grating. Food Environ Virol 5:52–60

Wasley A, Samandari T, Bell BP (2005) Incidence of hepatitis a in the united states in the era of vaccination. JAMA 294(2):194–201. doi:10.1001/jama.294.2.194

Weltman AC, Bennett NM, Ackman DA, Misage JH, Campana JJ, Fine LS, Doniger AS, Balzano GJ, Birkhead GS (1996) An outbreak of hepatitis A associated with a bakery, New York, 1994: the 1968 'West branch, Michigan' outbreak repeated. Epidemiol Infect 117(2):333–341

Wheeler C, Vogt TM, Armstrong GL, Vaughan G, Weltman A, Nainan OV, Dato V, Xia G, Waller K, Amon J, Lee TM, Highbaugh-Battle A, Hembree C, Evenson S, Ruta MA, Williams IT, Fiore AE, Bell BP (2005) An outbreak of hepatitis A associated with green onions. N Engl J Med 353(9):890–897

Yezli S, Otter J (2011) Minimum infective dose of the major human respiratory and enteric viruses transmitted through food and the environment. Food Environ Virol 3(1):1–30. doi:10.1007/s12560-011-9056-7

Yoneyama T, Kiyohara T, Shimasaki N, Kobayashi G, Ota Y, Notomi T, Totsuka A, Wakita T (2007) Rapid and real-time detection of hepatitis A virus by reverse transcription loop-mediated isothermal amplification assay. J Virol Methods 145(2):162–168

# Index

G. Sánchez, *Hepatitis A Virus in Food: Detection and Inactivation Methods*,                    49
SpringerBriefs in Food, Health, and Nutrition, DOI 10.1007/978-1-4614-7104-2,
© Gloria Sánchez 2013